12883790 X

THE CRAFT OF
STONEMASONRY

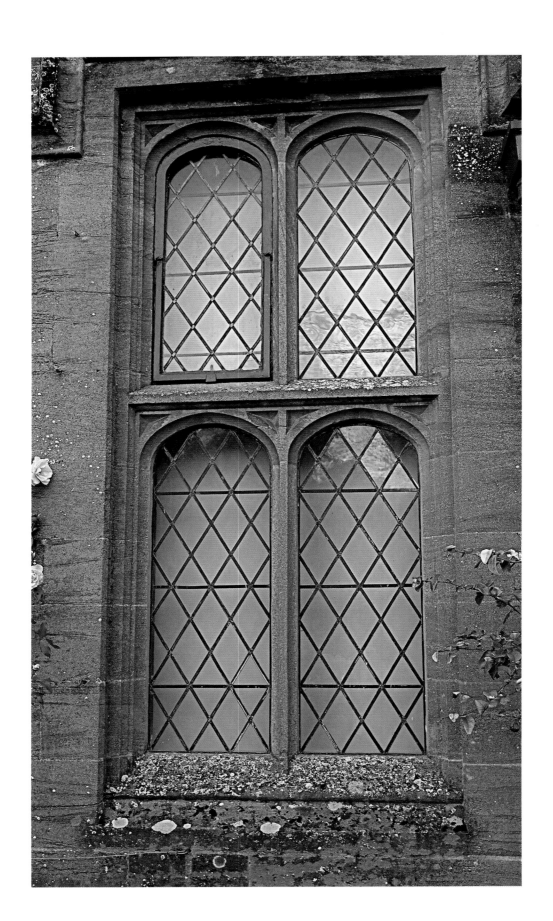

THE CRAFT OF
STONEMASONRY

CHRIS DANIELS

THE CROWOOD PRESS

First published in 2012 by
The Crowood Press Ltd
Ramsbury, Marlborough
Wiltshire SN8 2HR

www.crowood.com

British Library Cataloguing-in-Publication Data
A catalogue record for this book is available from the British Library.

ISBN 978 1 84797 385 6

Frontispiece: A Hamstone window with Tudor arches on a typical country house built of
stone in the south of England. Stone has practically defined the history of mankind and has
been the setting for some of the most momentous occasions.

Acknowledgements
Thanks for help and support are given to: Blossom, Lily, Rudi and Cadhla; Antony 'Hank'
Denman; Richard Mortimer, David Good and students at Weymouth College; Ian
Constantinides; Niall Finneran; Andrew Whittle, Ian Burgess and all other colleagues and
past students.

Typeset by Jean Cussons Typesetting, Diss, Norfolk
Printed and bound in China by Everbest Printing Co. Ltd.

CONTENTS

INTRODUCTION

Man's relationship with stone as a material to shape the world we live in effectively started when the first boulder was moved to block a cave entrance or bolster up the edge of a fire pit.

Here is a material that is available everywhere and in every size, from mountain down to hand size. For humans, the temptation to hit something with it would have been natural. Archaeological evidence shows that tools formed from stone were produced at the very dawn of *Homo sapiens* existence over two and a half million years ago. This heralded the dominance of man over nature by an ability to use tools to mould the environment to suit his needs – perhaps overtly self-destructive for the environment but certainly providential for stonemasonry.

Arguably man has utilized stone since prehistory as a component for dwellings, monuments and sculpture. Of the 'base' materials –earth, wood and stone – traditionally used in construction, stone is certainly the most enduring and (in our biased opinion) the most endearing.

Much of our understanding of how people lived and the culture they sustained is by interpretation of stone objects and edifices, where these are often the only remaining physical evidence of their existence. Just reckoning the considerable number of historic buildings that still exist is testimony to stone's durability and practicality.

As nomadic hunters did not need permanent structures, the first steps in building would be limited to placing blocks for fires, cairns for burials or blocking up caves. While a circle of stone found in Africa may date to 1,750,000BC, it was not until cultivation of crops effectively halted the nomadic lifestyle that man started to put roots down and began building

OPPOSITE: Victorian, neo-gothic doorway, showing the rich textures to be had from using stone. By use of this raw material it is possible to enrich the environment in which we live. The stonemason's part in all this is vital for machines cannot produce work to the same standard as the trained hand.

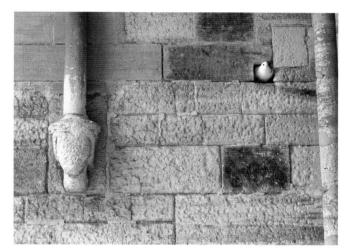

Coutance Cathedral.

Neolithic axe heads, Belgium.

Turkuza Maryam, Lalibela. Carved from volcanic tuff. (Photo: Niall Finneran)

ern counterparts pull the saw through, which allows for finer cuts and greater accuracy.

The committed spread of civilization had begun in Mesopotamia (the ancient city of Ur dates back to 7,000BC), and all manner of buildings started to appear in the region as the use of masonry rapidly spread across the continents. Elsewhere the Maltese, relatively isolated in the Mediterranean sea, were building huge elaborate temples by 4,000BC (a thousand years before Stonehenge) using only copper chisels and stone to work the material, just as the Egyptians did a millennia later to produce the pyramids. The centuries following produced the Greek, Chinese and Roman civilizations in all their glories.

The Spirit of Portland statue stands on the road leading to the Portland quarries, overlooking Chesil Beach.

in earnest. Jericho near the Dead Sea was probably the earliest farming community, and while the majority of houses were of earth, from about 11,000BC it was surrounded by a masonry wall, and there are records of a tower by 8,000BC.

The recent discovery in Japan of a massive ziggurat temple, 600 feet wide by 90 feet high in the waters off the coast, is believed to date to the same period, so the history of our craft fills out across the world. It is interesting to note that the western approach to handworking of stone (and wood) is almost the opposite of the Asian, in that westerners work away from the body and eject materials, whilst the style of the east is inclined to working into the zone of the artisan – which in many ways gives greater control of the process. The most noticeable example of this difference is that when sawing, western artisans push through the material whilst their east-

THE FIRST STONEWORKERS

Whilst the approach and their considerations on materials of those involved in architecture has been well recorded comparatively little is known about the lives of those who cut and worked stone in ancient times.

During the Egyptian and other Middle Eastern dynasties the stoneworkers were almost certainly no more than slaves. The same may be true of the marble cutters constructing the Parthenon in Greece or the Etruscan temples. The spread of stonemasonry was helped by the Roman military, for the legionnaires, the formidable fighting forces of the Caesars, were trained in all manner of construction. They disseminated their skills in the use of stone for magnificent buildings and structures, influencing the architectural style of the conquered lands during the massive expansion of the Roman Empire.

In western Europe during the Dark Ages the building of singular stone structures had faded into nothing, although across Asia, South America and Africa, civilizations continued to produce wondrous buildings and monuments.

The stonemason and his practice came about in the Middle Ages of Europe during the great flowering of Gothic architecture – a style as much a result of the (re)discovered delight of geometry gleaned from Islamic scholarly works as it was the church showcasing its all encompassing influence on the world of man. The new styles came about as a development from the Moorish influence and the translations of works on geometry from Arabic of the first millennium. Along with the Italian and Greek influence in the classical styles much beloved of western governments and institutions, these styles have defined our traditional architecture for over a thousand years.

The defining link between all the disparate styles is that they are (or strive to appear to be) constructed of stone. That is where our story begins – for without the humble stonemason these buildings would not exist.

Freemasons

The medieval stonemason would start as an apprentice to a skilled member of the guild of freemasons. The term 'free' does not describe their personal liberty but the type of stone used – freestone was a generic term for stone that could be

Roman column outside York Minster. Note only the capital and base are of worked stone, while the shaft is rendered to resemble stone.

A stonemasonry student being filmed for a documentary in the role of a medieval stonemason at Abbaye de Hambye.

Stonemason depicted on the staff of the Stonemason's Guild carried by members through Ghent in medieval times.

An apprentice piece by a member of the Compagnons du Tour de France, showing a stylized symbol of the Masonic sign.

Sculptures from the Parthenon, Athens.

cut in all directions, allowing complicated design and detail. The apprentice would be registered in the Masons Lodge attached to the particular building he was working on. This lodge was presided over by a master mason, who would oversee the quality of the training and work of his masons. Historically this is the origin of the present-day Freemasonry society, where they use the terminology of the stonemasonry profession, such as lodges, leather aprons and tools for their uniform and symbols.

THE ROLE OF THE STONEMASON

In medieval times, before the rise of individuals who designed without participating, the master mason would have acted as architect, construction manager and accountant. He was responsible for the whole construction project – designing, managing and recruiting all the trades and crafts to complete it.

Today architects will design a building down to the last tile and window on computer and then paper, but from there on, all aspects of the constructional stonework is still the province of the stonemason.

Artisan or Artist?

Before the renaissance, the concept of the individual artist did not exist, yet some of the greatest sculpture and decorative carving known to man is in the mainly religious buildings of the classical periods and middle ages. It is not our purpose to discuss here the nature of art, but what we can say is that

Medieval statues, West Front, Exeter Cathedral.

Stone heads for label stops, showing typical medieval colouring.

it was the stonemasons who created every aspect of these wonders.

It was a natural development that the spaces on a temple or cathedral should be used for adding emphasis to the power of God(s).

These medieval buildings were designed by master masons who understood what the client wanted as well as what they could get out of the stone – and their workforce. The great west fronts of the cathedrals were therefore carved with myriad saints, angels and depictions of the afterlife, to be viewed by the (illiterate) populace as stories and spectacles.

All this stonework would have been painted (and the painter of a statue was paid more than the carver of the piece!). Choirs and musicians would have been hidden in the wall, singing out from behind the vividly coloured statues and masonry. To a populace full of suspicion and belief in magic the effect would have been astounding, in an age devoid of technol-ogy. Our conception of these buildings is completely different from those who lived when they were raised. In many ways the existing cathedrals, churches and public buildings are but pale shadows of their past.

HOW STONEMASONS WORK

The Design

The individual pieces of stone (components) have to be identi-fied in the design and then converted into working drawings, templets and moulds. This might be done in the setting-out room by artisans – usually a talented member of the work-shop possibly working their way up to master mason. It might also be done in a small workshop by the person who will

LEFT: A young lad sawing stone with a frigbob stone saw, with help from his parents.

BELOW: Sawing excess stone off a boss stone in the college workshop. Whether young or old, the working of stone is a source of fascination for all.

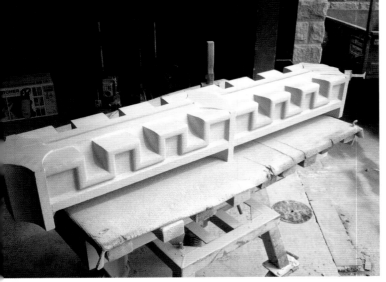

Detailed stone ready to be fixed. (Photo: Dave Edgar)

Mediaeval memento: a blind tracery piece carved in granite, the remains of an abbey in Britanny.

be carrying out the masonry work. Indeed a stonemason should be able to take the idea of a worked stone and carry out all the operations through to the finished item as part of a structure.

Getting the Stone

Getting the stone out of the ground is a toil that has many variations, all of which we shall lump under soubriquet quarrying. Once the stone is extracted the process begins by the quarrymen splitting it into manageable blocks. These are removed to the sawyers, who use machine driven saws to get the dimensioned blocks ready for the stonemason, though in less developed conditions pit saws and frig-bobs are still used on the softer stone.

When practical, much of the shaping of blocks would be generally carried out at the quarry to reduce transport costs; the debris would be utilized as core fill and, if limestone, burnt for mortar production.

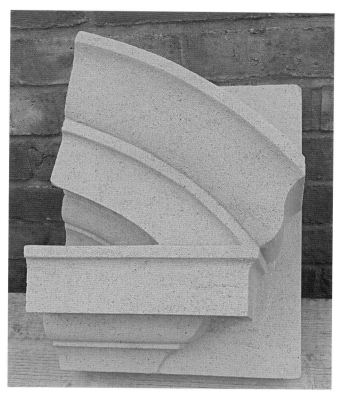

Springing stone for segmental tympanum – an advanced piece of masonry that is a combination of the the techniques explained in this book.

THE BRANCHES OF STONEMASONRY

Banker mason

The accepted term for those who use hand tools to shape

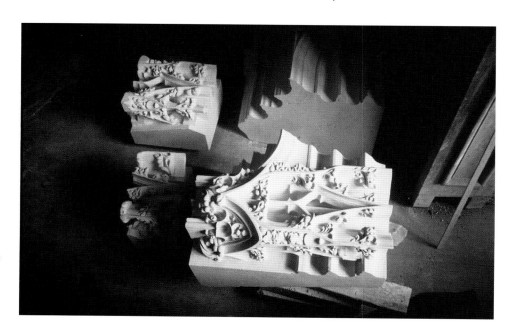

Carved and decorated pieces ready for insertion. (Photo: Ian Constantinides Ltd)

TOP: *Fettling in the components of new window. The slight variations where the moulding does not meet exactly are dragged in, using masonry tools and straight edges.*

MIDDLE: *Masons in Provence fixing new stone into the ashlar wall.*

BOTTOM: *Students practising fixing ashlar blocks at the college workshop.*

stone in the workshop is banker mason (traditional names include freemason, journeyman, tailleur de Pierre and marmiste). They would be involved in producing for use in construction the worked pieces of stone that tend not to have any decorative embellishment (unless it is a simple one, such as egg and dart cut into ovolo moulding).

Stonecarver

The enrichment of stone elements in buildings (such as capitals and corbels) is carried out by (architectural) stonecarvers. Carvers are considered a separate branch of stonemasonry, though often it is a progression taken up by a banker mason who shows talent in this area. Usually until they are selling enough work to do carving full time, incipient stonecarvers will also do a lot of banker-work.

Fixer Mason

The fixing (placing and building) of stone is carried out by the fixer masons. They should be experienced and competent enough to produce carved work to the same standards as a banker mason, as when fixing stone into position they may need to alter or finish off the pieces to ensure perfect work. Fixers are also usually recruited from aspiring denizens of the stonemasonry workshop, though much general building with stone is carried out by builders who term themselves masons.

LEARNING THE CRAFT

In the hierarchy of human skills stonemasonry deserves

its well-earned respect, providing, as it still does, the most durable buildings and monuments built of unprocessed materials. This has given substantial benefits over living in temporary structures and building short-lived monuments from wood or other degradable materials. It also, perhaps with foresight, gives us the opportunity of using our hands to create, without the recourse to machinery.

The underlying tenet of this book is that you the reader have a desire to become familiar with the skills of an artisan – someone who is practically independent and self sufficient, definitely not some cog in the machine of modern construction.

We will cover the aspects of work that an apprentice stonemason would be expected to be competent in: setting out, working and fixing. Instruction will be given in a relaxed and enjoyable manner using real examples from my own professional experience as well as covering many of the pieces needed by an apprentice to gain a technical qualification.

This learning process will take in many points of interest and show avenues of future exploration. As there are many varied stones in a building, so there are many components in the stonemason's formation. A building can have deficient or missing blocks yet still stand, just as a trainee can proceed to producing worked stone without experiencing all the tricks of the trade; the building would need to be repaired before it can be extended successfully, and the stonemason wanting to progress will need to occasionally go back to fill in the cracks or bolster basics.

The approach will encompass the approach, intelligence and enjoyment of beauty as well as mundane tool skills and workshop practice. There will be some digressions to help you understand and appreciate aspects of this profession that are often missed out in the modern educational system. This book will be distinctive from the stiffer, more formal, books produced in earlier times – but please don't discount these, either, as they are also useful.

Photographs and illustrations used to illustrate the text throughout. As stonemasonry is about techniques and skills, in some cases the illustration shows a described technique being used on a piece that is different from the one being taught. It is the skills that matter here, not the piece, so if you are of a mind to use this book without doing the set pieces, you will learn to realize that every job and stone just requires the correct approach.

The majority of the photographs used are digital, and certain features such as tooling or shadows have sometimes been emphasized by manipulating the image, possibly to the detriment of the quality of the image. The reason is this: without the use of studio equipment and lighting the nature of freestone can often give bland photographs but, as the majority of the pieces are actual work in progress, the presence of delicate stands and shields would have hampered work.

Read on and enjoy taking on the challenge of joining a company that reaches back to the dawn of history, as well as the pleasure and satisfaction of contributing to the heritage of our future in a fitting manner.

Fan vaulting, Sherborne Abbey.

CHAPTER 1

WORKING WITH STONE

The (natural) formation of stone provides the material in such a way that the general methods of its use in stone masonry construction have remained basically unchanged since the erection of the simplest cairns.

Bonding techniques are the most efficient way of building with stone, as evident from the great gothic cathedrals and castles through to modern block construction. Today's methods of cladding, in contrast, are merely economic and lightweight developments, preceded by many examples in history to give the impression of solid masonry. The extraction of dimension stone and processing it into usable components was – and still is, for all our technological advantages over the ancients – labour intensive, with finishing and carved work normally carried out by hand, underlining the relevance of this book.

The majority of masonry construction in the vernacular was traditionally rubble style of undecorated finish, using stone relatively unprocessed apart from knocking the odd corner off.

The resulting walls would be covered over with render or limewash – surface finish methods known as stucco, which were, and still are, used to give the impression of worked stone in more important buildings.

Occasionally though, the skilled hands of the stonemason working on these humble buildings would produce more elaborate work that was previously restricted to buildings of importance and wealth. In the Atlantic fishing town of Roscoff,

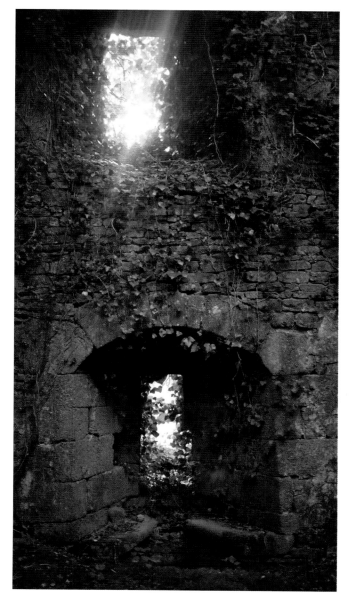

Ruins, Northern France.

OPPOSITE: *Part of an old Roman wall using igneous stone and sandstone, both indigenous to the local area. It is this huge variety of material ready to be plucked from the ground that gives a local quality in buildings of a traditional style; the repair and maintenance of historic buildings relies on an awareness of the material and its source.*

Flint and brick walls with stone quoins, Wiltshire.

for example, many buildings have elaborate stonework depicting the life of the sea, and it is reasonable to assume that fishermen would have turned their hand to masonry when the seas were too tempestuous.

With increasing levels of industrialization, extraction from quarries and the production of dimension stone became more economical. This allowed the use of worked stone in buildings of lesser stature to become more commonplace and widespread, as materials were transported further away from their usual region of use, creating a greater demand for the skills of stonemasons. The extensive use of French stones quarried near channel ports used for buildings in the south of England during medieval times was not just the choice of the Norman stonemasons, who knew their stone, but it was also expedient for transport by water.

Fake jamb and voussoirs, Exeter. Moulded out of render to look like stone.

HISTORICAL CUTTING EDGE

It is noteworthy that the hand tools used by stonemasons working today are practically identical in design to the ones used by their predecessors since time immemorial.

ABOVE: *Chisel blades: (left) a pitcher or handset; (top) a tungsten carbide blade; (right) fire sharpened boaster.*

LEFT: *Old French mallet and chisels on display at a Compagnons du Tour de France exhibition, Provence.*

There has been considerable advance in the materials used for the manufacture of these tools, however, from the hand held hammer stone of the Neolithic age. Although the Bronze Age only allowed the use of relatively soft copper alloys, stonemasons were still able to construct massive stone structures – and constructed well. The downside of having these soft metals for cutting would be the short working life before sharpening was needed again.

The turning point in efficiency terms was the discovery and adoption of tempered (fire sharpened) iron and steel. For chisels this meant that cutting edges would remain sharper longer and thus increase productivity.

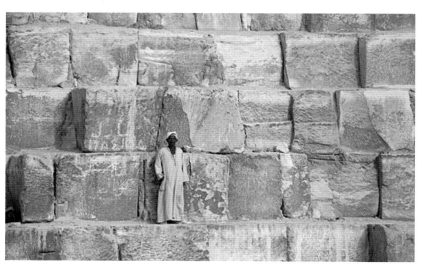

Nowadays the use of tungsten carbide as a cutting edge for many stonemasonry tools offers even greater increases in the amount of work possible before needing sharpening. Interestingly given the choice, many stonemasons still use fire-sharpened tools, preferring them for numerous applications.

While the great Pyramid of Cheops contains 2.5 million cubic metres of stone, one of many such wonders erected by the Egyptians, the amount of stone used in the French medieval building period was actually more than all Egyptian construction combined.

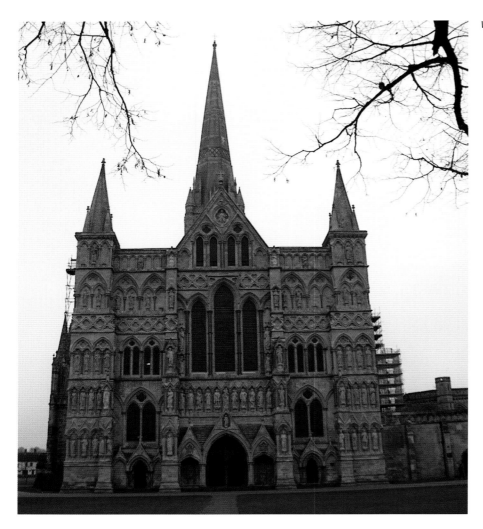
West Front of Salisbury Cathedral.

Stacks of Garston stone in the quarry, Canada, showing bed heights of stone straight from the ground.

The incredibly wide variety of stone types available means that there can be a great wealth of textures and colours used on the exterior of buildings. This can pose problems when replacement stone for an historic building has to be found, as it is not uncommon for original quarries to be worked out or closed down; tracing the origins of a type of stone can sometimes involve an amount of detective work.

Stone (or components made to look like stone) will always have a use in the majority of important buildings. Without its existence many (if not most) of our cities and towns would not be such visible and accessible records of our cultural heritage.

Consider: Bath in southern England, with its fine Georgian terraces and imposing Prior Park built of warm yellow-grey limestone from local quarries, is probably one of the most important towns in the world with regard to cultural significance due to its architecture and stone.

PORTLAND STONE

Westminster Palace, London. The stone for this was originally Anston (magnesium limestone), but unfortunately the stone was not up to the job and reacted badly to coal burning pollution, so it was replaced in the 1920s with a limestone – Clipsham. In the

seventeenth century many of London's great buildings were destroyed by the infamous Great Fire; the massive rebuilding programme used Portland stone, transported to London by barge from the Isle of Portland.

The stone from Portland was loaded by crane off the cliffs directly onto barges waiting in the turbulent waters, before being transported around the coast to site.

Stone was used to construct locks and gates on canals, subsequently allowing the movement of material across country.

Venice as a floating city is built of stone that reflects its sea-faring nature, so although it does not have a local material, the presence of the Adriatic Sea allowed it to import Istrian limestone.

The great Egyptian pyramids and tomb complexes show the result to which the massed resources of a despotic nation and the transport possibilities of the Nile could be put.

CONSTRUCTION METHODS

'Solid Wall'

It is not often the case nowadays that buildings are con-structed using stone as the primary structural fabric, other than as facing or cladding. Traditionally built buildings with ashlar faces (squared stone laid in regular height courses) however, tend to be solid wall construction falling into one of two types, *Opus Incertum* and *Opus Testaceum*.

Massive walling – in castles, churches and other great houses – is composed of two skins of stone, with a rubble core sometimes metres thick. This style carries the Latin name of opus incertum.

Interestingly there can also be timber or metal inserted within the wall as ties for bracing and bonding. Due to the size and construction of this type, structural problems can often be as significant as the deterioration problems of the stone itself.

It stands to reason that the need for the movement of

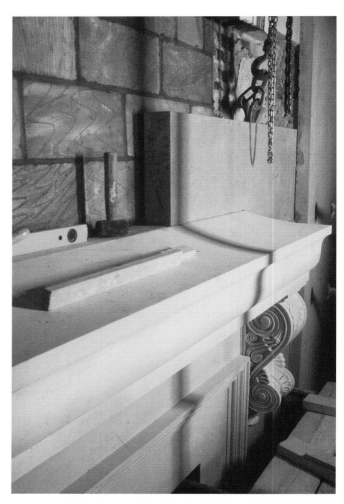

New stone. British Museum built in the modern style with a cavity and tied back to the concrete blockwork. Note blocks being lifted in by lewis. (Photo: Ian Constantinides Ltd)

Massive walling for fortification in Dorset. The facing stone was built at the same time as the core.

have a skin of stone on a brick or rubble backing. Once again we acknowledge the lineage of this work from ancient times with a Latin moniker: opus testaceum.

people around important buildings without disturbing those of importance would require hidden throughways and discrete openings that would be incorporated in the structure. I like to think that there could be the proverbial secret passage waiting to be discovered by the lucky stonemason replacing a few loose stones.

Lighter walls can have two skins of masonry relatively close together, often with stones bonded through front to back, or

Cladding

Cladding, using thin slabs of stone hung on steel or pre-cast concrete framing is now the norm for a 'stone' building.

The use of cladding is not entirely a modern process. Many important historic buildings have just a thin skin of stone over a rubble or brick core; this is known as opus reticulatum. There are typical historic and modern examples of keeping the cost down by minimizing the amount of worked stone required.

Buildings prior to the late nineteenth century tend to have a solid construction of some cheaper or more readily available material, whereas the next generation used frameworks of iron to support the stone. These modular buildings rely on machinery such as frame saws to produce consistent dimensioned stone components that can still exhibit some of the problems that are found in solid construction.

Granite gatepost, France. The hard stone is simply worked. Note the ugly modern mortar and the poor application of it.

Highcliffe Castle, Dorset. Steel is inserted to support and attach new stone to. (Photo: Ian Constantinides Ltd)

As most buildings, especially vernacular, used materials and labour that was indigenous to an area, then styles and methods of building can often exhibit idiomatic trends that should be understood when repair is necessary to them; it is wise to approach each building as a unique object.

Stone can be split for use as roofing. Known as stone slates, these are actually limestone, which cleaves well into thin slabs. The style here is of diminishing courses, which uses larger stones first, decreasing in size to the apex.

CLASSIFICATION OF STONES

The primary classification of stones (used in construction as well as in science) is by their geological origins – that is, the manner in which minerals and materials are brought together, amalgamated or altered are principal factors dictating type and ultimately their usage.

Some stones, like granite or flint, are hard to work and will be usually used in the state that they are found or lightly worked.

Slate, and similar, is dense and splits into flat plates and can have its edges squared off to be used as roofing or paving.

Farm building, Brittany. The wall is constructed from massive sheets of slate, stood on edge.

Monument work requires stone that will carve and polish well, like marble and alabaster.

Most areas will have originally utilized the local materials available as best they could, adapting the style to suit the stone, or using it minimally – such as plinths for earthen and timber framed buildings or as a rubble background for applied finishes. The geological situation of an area is habitually the deciding factor on what stone the buildings were made from and how they are constructed. It is remarkable to consider that (geological) processes are under way all the time, and the formation of stone continues with contributions from every scraped boot on a footpath as much as the dramatic eruption of volcanic suppurations.

In this book we will be using limestone and sandstone – material extracted in uniform height blocks and worked with hand tools. Despite the presumption that the working and shaping of the stone in the exercises is carried out with sedimentary stone, however, it is essential for a stonemason to have a good knowledge of the varieties of stone – they are everywhere and can crop up in the most unexpected places. (Note: geologists prefer to use the term rock rather than stone, but here we will skip between the terms as appropriate.)

All stone is composed of minerals in various forms and quantities. The geological classification of the stone is determined by these minerals – how they came to be together and what processes were involved in their and the stone's formation. Petrology describes the history and formation of the stone, and petrography describes the physical and chemical structure of the stone.

Planet Earth can best be described as a large ball of molten material called magma (lava) that has a thin (in proportion) skin or crust composed of cooled, solid material – the bulk of which is known as igneous rock.

Igneous Rock

Igneous rock can be either extrusive (basic) or intrusive (acidic).

Extrusive igneous rock is formed by molten magma being ejected (volcanically) onto the surface and cooling.

Intrusive igneous rock is formed underground by magma collecting into huge masses (batholiths) pushing into surrounding rock; as time goes by and through subsequent geological processes these can become exposed on or near the surface and quarried. South west England and mid Canada are situated on batholiths of granite that have been exposed

as the covering material is eroded away; one striking feature of this the creation of the moorlands studded with tors.

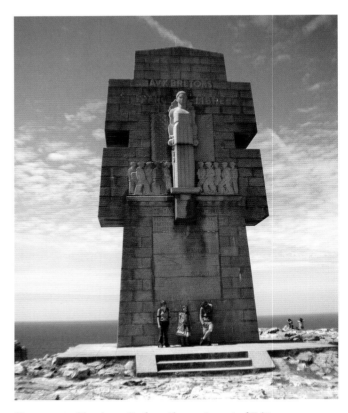

Monument of local granite from the west coast of Brittany.

Durdle Door, between Purbeck and Portland, is limestone that through geological disturbance has been turned on edge, allowing weaker stones to be eroded away.

Sedimentary Rocks

Igneous rock constitutes most (95 per cent) of the Earth's crust, yet is only a quarter of the material available on the surface. The sedimentary rocks, on the other hand, are only a twentieth of the crust material but make up the rest of the surface material. The younger sedimentary rocks will tend to cover the bulk of the older igneous rocks. These relatively younger rocks can be generally described as having igneous rock material as a common origin but not necessarily in an obvious or direct manner.

As the name suggests these stones are formed as sediment (from whichever source) identified in the wild by the layers of varying type.

Cliff of sedimentary stone, showing how more durable beds weather less.

Sedimentary stones are classified by considering more immediate (in geological terms) sources, which better explains their make-up.

1. Volcanic ejectimenta.
2. Mechanically disintegrated rock fragments.
3. Products of rock decomposition.
4. Precipitates from aqueous solution.
5. Organically derived.

This list may well be considered a simplification, as a stone formed in one way is often affected by the formation of other(s) and can create a completely new material.

Volcanic ejectimenta

Sedimentary rock that is formed at the surface will in most cases have traces of material of a volcanic origin, coming

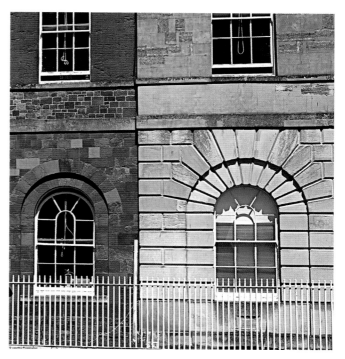

Walls of various stone, Exeter Castle. Local sandstone and basalt (left), and rusticated ashlar of Portland stone (right).

from violent activity when large amounts of pulverized lava is thrown up to form beds similar to sand and other clastic materials. These are known as tuff (often incorrectly called tufa) and ash.

Closer to the vent heavier, coarser pyroclastic rocks (agglomerates and breccias) are formed over vast periods of time. As the process disrupts the surrounding country it may dislodge or envelop fragments from other types of stone, so you can end up with sedimentary rock inclusions.

Mechanical disintegration

Rock on the planet crust is broken down into fragments by the effect of erosion and weathering through glacial movement, water and weather.

These fragments of rock are always in motion, and through friction they become smaller and more uniform over time. A first-rate example of this can be found in the south of England at Chesil Beach, a long pebble beach that shows an intermediate stage in the erosion of fragments. It is always under constant change through movement of the sea, where violent storms can lift and redeposit millions of tonnes of pebbles in a single night. The grading of the stones here is so marked that some locals claim to know where they are standing on the beach by the size of the pebbles.

When these particles have been ground much finer, they are easily carried in water and deposited in situations such as mouths of rivers, where they will build up, becoming compacted and cemented together by pressure and/or from minerals leaching into them, leading to the formation of sedimentary stone.

CLASTIC FRAGMENT CLASSIFICATION

Clastic stone has particles of itself or other materials in the matrix. Constituent fragments in a clastic stone are classified according to the mineral nature of the fragments and also by the size of these fragments.

The sizes below are useful when identifying stone as well as analyzing or designing repair mortars.

FRAGMENT CLASSIFICATION

Name of fragment	Diameter (mm)
Boulder	256 or more
Cobble	64–256
Pebble	4–64
Granule	2–4
Sand	0.062–2
Silt	0.004–0.062
Clay	< 0.004

Conglomerate stone, Devon, showing how the inclusions give it an almost artificial appearance.

Rock decomposition

The chemical decomposition of rock is a highly motive force that accounts for approximately 70 per cent of the mass of sedimentary rock. This decomposition can come about in many ways:

- Water (often under great pressure and high temperature) containing dissolved oxygen and carbon dioxide will start the breakdown of igneous rock. As this happens other alkaline and highly reactive compounds are often formed; these caustic solutions can also dissolve silica and silicates at normal temperatures and pressures.
- The oxidation of sulphides forms sulphuric acid, once again highly corrosive and effective in breaking down many rock-forming minerals, as can organic residue form acids that play a significant part in the breakdown of rock.
- When a crystalline rock has been completely weathered the resulting products can include:

Soluble salts
Colloidal substances
Insoluble secondary minerals
Chemically resistant minerals

All of these are transported in different ways and go through various stages to form different kinds of sediment.

Precipitates

When minerals such as calcium are taken into solution, there is a stage reached where, by oversaturation or through chemical reaction, the mineral solidifies out of the water, forming beds that over time can compact to a relatively solid mass. It can also leach into other minerals or the remnants of animals'

Calcite forming under an arch of Purbeck stone, Corfe Castle, from unset mortar put in during conservation work – caused by lack of understanding of the practical application of traditional mortars. Note the soulless modern attempt at representing rubble infill.

existence – such as skeletons or shells, that are already deposited as a result of warm seas or other situations.

Under the surface of the earth, pressure can create superheated water capable of dissolving a great number of minerals and carrying them into different rock formations (you can start to see the links and crossovers between types of stone).

Organically derived

Living organisms extract substances dissolved in the waters of the earth, the most important being calcium carbonate used in the shells of many invertebrates. When conditions and amounts are suitable (over millions of years) these materials accumulate to form substantial deposits. Due to the great number of creatures (and time involved) other materials, though only present in small amounts, can also form significant deposits such as calcium phosphate and silica (the main inorganic substance in many sponges). Mineral oils and gases found in various sediments are formed from the diagenetic processes affecting the softer tissues and fluids of organisms.

Plant tissue, composed of carbon and other minerals, accumulates to form carbonaceous sediments – the first stage of this is peat, which leads to coal and oil.

In some limestones there is occasionally a presence of ore minerals, far removed from the normal areas of mineralization, which can be attributed to the small amounts of metals found naturally in invertebrates; that these are significant is due once again to the numbers and time involved.

ACCURATE IDENTIFICATION

To successfully identify a stone all factors – deposition, mineral and chemical composition and so on – must be taken into account for accurate classification. Try to get away from the general terms sandstone or limestone.

While this may look like quite a specialized area generally attributed to geologists, whenever appropriate stone should be replaced like for like, so for the stonemason this knowledge is always useful. Accurate identification is also important to conservators involved in the decay and preservation of stone at a technical level beyond replacement.

Bowers Quarry, Portland. Beds of limestone are visible at the edge of the working. A mine entrance is to the right.

Metamorphic Rocks

Due to the often immense periods of time since the processes of rock formation started happening it is not uncommon to find sedimentary stones that have been broken down even further, deposited in different situations, with further additions or alterations to their mineralogy.

After igneous or sedimentary rock is formed and settled into a comfortable petrographic niche, through geological movement and change it may be affected by great pressure or heat processes that bring about the formation of another type of rock – metamorphic.

The great weight of the overlying strata compressing the rock or contact with the intense heat generated by the molten magma intruding into an area of rock can bring about changes to the mineralogy and structure of the rock. This metamorphism may result in a range of stones from marbles that are very similar to limestone through to completely different materials that bear no resemblance to the parent rock.

Some common examples are:

    Limestone to marble
    Shale to slate
    Sandstone to quartzite
    Plant remains to coal.

STONE SELECTION

There are many considerations when choosing a particular stone for use in new construction, replacement or repair. Obviously when the stone is to be a replacement in an historic building it is preferable for the original type of stone to be used. Often it may only be possible to effect a compromise by using stone from a non-original source to match the original stone as closely as possible.

Quarries become worked out, closed or even lost. The majority of smaller quarries were usually opened up for local building, meaning there may have been only a limited amount of the stone available, which has subsequently been used up.

It is important to note that stone as a natural product can vary greatly in the quarry and even in the same bed. This

David, *by Michelangelo – artistry in marble.*

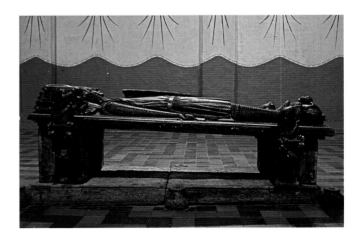

Tomb top, Belgian black marble, Ghent.

CLASSIFICATION OF STONE

Igneous	Formed from molten magma in the Earth's mantle, as igneous intrusions in the Earth's crust and lava when extruded by volcanic activity	Granite Basalt Serpentine Pumice
Sedimentary	Formed by the deposition of material in layers on lake/sea bed or at river mouths, as the result of erosion, precipitation of minerals or accumulation of animal skeletons	Limestone Sandstone Chalk Ironstone
Metamorphic	Stone that has deformed and recrystallized to varying degrees by being subjected to great heat and/or pressure	Marble Slate Gneiss

Portland stone was used to replace the softer Beer stone. This was due to the scarcity of the original material and also because of the superior weathering performance of Portland.

Beer Quarry, Devon.

means that even if the original quarry is still producing stone that has the same classification and name as the original it can be different in performance, texture and colour.

Whenever you read promotional material for stone or order from a company brochure, there will be a small print disclaimer; stating 'stone is a natural material and may vary in colour and quality from the samples shown'. This is important to know if you have very specific requirements, as the stone for sale could be very different from what was expected. You may need to visit the supplier and pick out suitable stone.

Replacement stone must perform to the same standards

as the original. This entails recognition of all the qualities required by a stone and how the replacement compares to the original in a standard array of criteria. Luckily there are standards and codes of practice that cover stone and its performance criteria; all stone quarried today for use in construction will have need to have a data sheet covering these.

Identification

Some stones can be straightforwardly identified in situ by the experienced eye; others require a knowledge of architectural

FOR REFERENCE

There are stone collections held by institutions such as the Natural History Museum in London with some 20,000-plus specimens of stone that can be approached for help in the hunt. With this number of samples it is important to have a good working knowledge of stone to narrow down the search.

All major UK stone suppliers are members of the Stone Federation of Great Britain, which can often be the first and last source of information regarding the obtaining of stone. The Marble Institute of America and Australian Stone Advisory Association are their country's respective organizations.

Selection of building stones, supplied by various quarries and sawyers. It is useful to apply to local suppliers for samples to keep as reference.

and constructional history that can be used to check against recorded or known contemporary use in similar applications. It may however be necessary to compare the stone to a set of reference samples. This can be as simple as holding it against a photographic reference or as complicated as producing a petrographic thin section slide and identifying the constituent minerals by polarized light analysis or micro-chemical test (this level of identification is best carried out by trained conservators or geologists).

The historical development and trading methods of a coun-

SANDSTONE OR LIMESTONE?

As mentioned most people identify sedimentary stone as sandstone. In truth it can be difficult to distinguish between the two, especially if they are firmly attached to a building. Here are two methods that are simple and while not foolproof will give you a head-start.

The first works best on old stone, preferably affected by pollution, and can even be done from photographs. Look at the way the stone is stained: if the upper, rain washed, surfaces are stained it will be sandstone, the discoloration due to oxides in the material reacting with (acidic) rain – in essence it's rusty.

If the stone appears clean on upper surfaces and dirty – or has black crusty build-up – on the sheltered undersides, it is limestone; the (acidic) rain washes out the calcium which converts by reaction to sulphate which, being water soluble, is usually limited to protected areas.

The second method is very easy, but only identifies limestone. Scrape a tiny area of the stone clean, blow the dust away and put a drop of acidic liquid* on the surface; if it foams it contains calcium carbonate – the binder of limestone.

RIGHT: *Badly stained limestone doorway, Ghent. Note the rain-washed areas are clear of staining.*

LEFT: *The acid test. Similar looking stones: the piece on the left is sandstone – no reaction; the piece on the right reacts, showing it is limestone.*

*In my professional capacity I am allowed to carry hydrochloric acid, which is highly hazardous and not recommended; a milder acid such as vinegar will give a reaction that may need to be viewed by magnification.

try can influence the use of stone. Great Britain, formerly a major sea-faring race, has imported many stones as ballast in ships, as well as 'non-native' stone such as Caen from Normandy, introduced following the Norman Invasion. With the introduction of Caen came French masons familiar with the stone, who also gave us much of our terminology.

Colour and Texture

Colour and texture are self evident, but be wary as what you see may not be an indication of the final condition of the stone as colour can change when dry or from the effect of

Similar looking stones are not always the same.

weathering. Inspection of abandoned faces in the quarry can give some indication of colour change. It is helpful if possible, to observe buildings constructed of the same stone.

There will be a problem with repair, in that new stone pieced in can be substantially different in colour and often texture. This is due to the effects of pollution, decomposition and past cleaning leading to the original stone having a greatly altered appearance – generally more noticeable with sandstone than others.

The right side of this stone has a roof sheltering it from the effects of weather, which results in differential staining patterns.

Chilmark stone in a church tower, Dorset, the colours showing a checkerboard of different reactions to weathering by stone from the same quarry source.

It is a quite common mistake to use a stone for repair or replacement purely because it has the same colour. This can go wrong, as over time it will invariably weather and decompose differently, with often strikingly bizarre results.

New stone of the same type as the original.

The texture of a stone is due to the minerals it is composed of, its pore structure and its manner of deposition. Evenness of texture and size of grains are not to be taken as an indicator of physical strength or durability.

Durability

The durability of a stone will usually be affected by many factors. As these will be different for every usage (and often the stone itself will not have a uniform composition or structure) durability tests should not be considered as having universal advocacy.

Purbeck marble (actually a limestone) splitting as moisture washes out binder.

THE SALT TEST FOR DURABILITY

The standard durability test is by the crystallization of a soluble salt (sodium sulphate) from a solution the stone is soaked with. The expansion of the crystals within the stone, over a number of cycles, will break down the material of the stone; the result is expressed as a percentage of the stone compared to an industrial standard. Widely regarded as the benchmark for assessing exterior durability of stone, it gives results that are best described as 'empirical' when considering the makeup of stone and the uses it is put to, which must also be considered.

Sulphation test showing the destructive effect of salts being sucked up through stone, and evaporating.

As a general rule, the coarser the texture of a stone the more durable it tends to be. Good examples are English Portland Roach and Pont du Gard of France; conversely while Hamhill stone is coarse textured and very durable, it can have poorly cemented beds and inclusions of clay that weather out quickly and lower its strength.

Depth on Bed

Sedimentary building stone size is limited in two ways: bed height and joints.

Bed height is the depth of a homogenous layer of stone as a result of a period of deposition, or where different materials

Block of bed height Hamhill stone.

Blocks of bed height Chilmark stone waiting to be used at Salisbury cathedral.

or geological factors inhibit the composition build-up of the stone layer. The stone between two of these planes is referred to as a bed. Obviously this dictates the maximum height of available stone for natural bedded masonry.

Joints, in quarry stone, are breaks or lines of cleavage perpendicular to the bedding plane. These are due to pressure or movement during the time since it was laid down and may limit the length of stone available, which would be crucial for lintels, for instance.

Stones that are quarried will provide usable dimension stone limited to what is usually obtainable in a particular size for that bed in the quarry, and when specifying or designing a piece it is important to know the dimensions available.

Availability

It is essential to ensure that there is sufficient quantity of a particular stone of the correct size that can be delivered within the necessary time – some stones are in great demand and need to be ordered well in advance. Stone that is out of production may still exist, sitting in the corner of some mason's yard perhaps, so if the amount required is limited it may, after some detective work, be possible to find a supply. Alternative stones will usually be offered, and these must only be used if they satisfy the demands of the project as well as being approved by all concerned parties.

There are cases where a stone needed is available but the methods used in quarrying, such as blasting, meant that

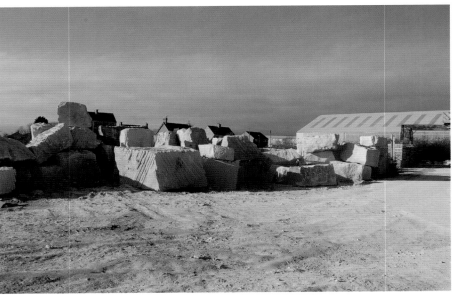

Blocks of Portland at the quarry prior to being processed.

RIGHT: Modern, misguided, attempt at repair, where stone is used in the wrong orientation (end bedded) in a window jamb; that this is so obvious is highlighted by the correct use of the same stone next to it in the jamb.

BELOW: This stunning garden in California that features stone from Somerset in the UK shows the advantages of a capable transport system in getting that 'perfect' material – having lots of money helps here! When the work is thought out carefully and craftsmen, architect and client work well together, the result is work such as this.

the majority of the stone is unusable; the disproportionate amount of wastage causes a large increase in the projected cost.

Cost

Cost is affected by transport, ease of working and wastage (which can be exacerbated due to irregular quarry blocks, high incidence of internal flaws or uneconomic detailing). The overall cost should be based on fixed stone and not comparative price per cubic measurement of un-worked stone.

Strength

Strength is not usually of the utmost importance as even the weakest of stones may be expected to withstand normal use loads when employed in the correct orientation. Abnormal stress concentrations such as uneven settlement or the expansion of ferrous cramps can cause even the strongest of stones to fail.

OUT OF THE EARTH

Quarrying

The two principal branches of the quarrying industry are dimension-stone and crushed-stone (aggregate).

In dimension stone, which concerns us, blocks of stone are extracted in different shapes and sizes for construction and monument production.

In the crushed-stone industry, granite, limestone, sandstone, or basaltic rocks are crushed for use principally as road stone or aggregate for the building industry. (This is what we need when making mortar.)

- Limestone (calcium carbonate) aggregate is used in farming to alter the pH of soils, and to produce calcium oxide (quicklime) for the iron and steel industry.
- Limestone and clay rocks are also used in the production of cements.
- Alabaster or calcium sulphate is quarried to produce gypsum, another major material in construction.

Belgian limestone – a very dense and durable stone commonly used for lintels and sills – was a complete failure here.

Methods of Extraction

Quarrying of dimension stone is carried out using all manner of methods and equipment – such as hand tools, explosives, thermal lance or power saws, channelling and wedging – according to the purpose for which the stone is extracted and also the level of technology and investment available.

Hand Tools

For quarrying stone that lies in easily accessible beds hand-tools alone may be used.

To begin with, a row of holes several centimetres apart is made with the drill and the hand hammer, partly through the layer, or stratum, perpendicular to its plane of stratification and along the line at which it is desired to break the stone. Each hole in a long row is filled with three wedges, shaped so that one may be driven down through the others, the method being known as plug and feathers. Each plug is struck with a sharp blow with a hammer in succession; the operation is repeated several times, and the combined expansive force of

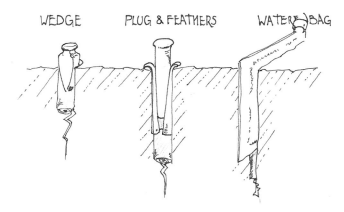

Methods of splitting stone.

the plugs and feathers finally becomes great enough to rupture the rock.

This method is also useful in the workshop for the splitting up of boulders into manageable blocks.

Industrial Methods

Explosives are commonly employed for detaching large blocks of stone. The drill holes are put down to the depth to which

it is required to break the rock; they are then partly filled with explosive and detonated.

Channelling is done by cutting long, narrow channels in rock to free the sides of large blocks of stone using self-propelling machines. Formerly steam driven, but now generally internal combustion or electric engines, they move a cutting edge back and forth on a rock bed in which the channel cut is to be made. Once the channel is deep enough to permit the insertion of wedges, the rock is split, the cut guiding the fracture. A modern approach is to use hydro-bags that are dropped into slots and expand by water pressure.

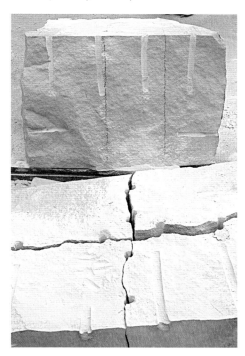

Block split using plug and feathers.

Channelling and wedging is extensively used in quarrying marble, sandstone, limestone, and the other softer rocks, but is not always successful for granite and other hard rock, these can require thermal lances using pressurized flame to gouge out in the rock.

Another method used in slate, granite, and limestone quarries uses the combination of a power saw, an abrasive and water as a lubricant and a coolant. The saw cuts a narrow channel, the primary or initial cut, which is then either expanded by a wedge or blasted.

Long continuous wires coated with diamond grit are used in the marble quarries of Carrera and some granite quarries; these cut straight edged blocks out of the quarry, a method that is useful for a rock that does not cleave readily.

Production of Sized Stone

The rough blocks of stone obtained from the quarry are chosen by size to be used for certain applications; their sizes are regulated by bed height and distance between joints. It is important to know these dimensions as they will affect the choice and availability of stone for building.

The other points to consider are the quantities and quality of stone available and being extracted. It is best to be aware that as some quarries keep prime quality or scarce stone in reserve you may have to do a bit of rooting about.

Sawing

The rough quarry blocks are transported to a sawmill where they are cut to size, first by the primary saw to obtain squared off sides and then by secondary sawing that whittles it down to the exact dimensions ready to be worked by the mason.

For cutting, saws have either discs with diamond dust embedded in the metal to provide a durable cutting edge, or long toothed blades (again using diamond as the abrasive) to produce uniform thickness stone (as in cladding) set together as a frame saw.

Quarry blocks straight out of the ground are surprisingly cheap, which is obvious because without specialized equipment they are awkward, if not impossible, to transport and process.

Most quarries have their own sawing equipment (and masonry shops) to produce the more profitable cut stone.

Sawing slabs to length.

This localized production is useful as stone can be inspected prior to and during processing. If you are going to be buying your blocks, the main objective will be to get it cut to the correct size for the job. It is worthwhile to seek out a reliable stone-yard or good sawyer and befriend them; badly sawn stone can add days of labour to the task, whereas accurately cut and squared material is worth travelling for.

FROM THE SAW

Sawyer's work should be appreciated, as competent and conscientious sawing, whilst producing stone to be carved, can do much of the work of cutting fillets and chamfers using the saw, thus saving the mason time and reducing costs.

Traversing with the saw to form a curve.

A block of stone cut to the external dimensions required is termed sawn six sides (SSS) – the common method of specifying stone. (Imperial and metric measurements become names for tools, so it should be no surprise that stone for masonry work and sold SSS is traditionally priced by the cubic foot – though measured in millimetres.

FAULTS OF STONE

If stone were completely impervious to erosion or decay then we would not exist, for our physical environment (or more grandly – life on earth) is primarily a product of the weathering of stone. The decay and conservation of stone is an enormous subject – it would need another volume to cover these

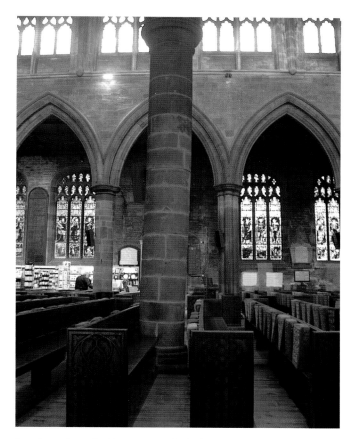

Leaning column in Gloucester Cathedral.

Lines and Vents

The most obvious faults are lines of weakness where if pressure is applied or the stone is knocked it will break in two or more pieces. These are hairline fractures and may not be visible at the surface, which makes it incredibly frustrating if the stone breaks during working.

The first check is to look at the stone carefully and, if you have any doubt, get an airline or hose and blow/wash the surface clean to remove dust and expose the fracture line. Alternatively hold a chisel loosely and tap the stone all over with the blunt end. The stone will have a definite ring, the volume and clarity varying across the types; if there is a vent the sound will be muffled and dull – distinctly different from a solid stone. Get into the habit of always tapping stone – you will soon become savvy.

Holes drilled for pinning across vent in stone. The dark marks at the lower part are resin spills.

– but here we will deal with the basic problems that can occur when working with stone.

Stone is a product of many sources and environments, and can have faults inherent in the block before it is worked on.

Natural vent in stone shows up after many years and the stone splits off.

A vent may not preclude the use of the stone if it does not affect the working area – it could be that the vent is starting off at the back and has not gone through the whole stone.

To prevent the spread of the vent or the stone falling apart, you can insert scissor pins. Simply drill holes across the line of fracture into sound stone, making them angled to each other so they 'cross' at the fracture (do not use percussive or hammer action as this could shake the stone apart). Clean the hole thoroughly of dust, using an airline if possible, or bottlebrush and wash out with water (adding some industrial alcohol to the water will improve the effect) and let them dry. Top up the holes with resin and insert stainless steel dowels to make a complete fill. Allow them all to set before working the stone.

Veins and Fossils

Stone can often have sections where the material differs greatly from the main body; these can include hard calcite veins, fossils and muddy inclusions.

Calcite or quartz veins will be the product of mineral movement through the vents of a stone and need not necessarily be a problem. Where faults can cause problems when working the stone is that they can be harder than the surrounding material, causing high spots where the chisel has a harder job getting through. This will also make rubbing and dragging surfaces difficult.

The problem with fossils is that they go through mineral change as well and can sometimes pop out, leaving an impression below the worked surface.

Inclusions

Some stones, in their formative years, will have picked up organic materials that have not mineralized, such as wood, which will have turned into a muddy spot within the stone; this will disintegrate when cleaning the stone or when exposed to the effects of weather. Usually these inclusions are not of any great extent – if they were the stone would have been rejected. They are unsightly rather than dangerous, though working up to one will require care to not spall off more stone.

They are often dark, and the runout will stain paler stones, so it is worth getting out all the (soft) material with a pressure washer or air-line at the soonest opportunity. They can be filled with a lime mortar using sand and stonedust to blend in with the original. When it has cured it can (very gently) have tooling applied to match. Two rules to observe here: never use cement and never leave the fill smooth.

Soft Beds

Occasionally sedimentary stones will not be all the same quality; they may have had periods of deposition resulting in much softer beds running through the extracted bed height. These will have been subject to differential weathering, resulting in lines of erosion that can be quite marked. Whilst working the stones does not show up these marks, the effect of weather and pollution can drastically reduce the lifespan of these stones. Experience of the type of stone used in certain appli-

cations and how they traditionally cope will help in choosing appropriate stone. It used to be said that a stone should never be used more than a hundred miles north of its quarry.

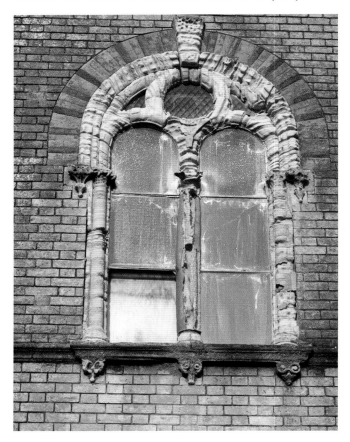

Bath stone is particularly susceptible to pollution. Note the end-bedded mullions (used this way to get the height) are sloughing off badly and their structural performance is compromised.

QUARRY CHECKLIST

A quarry is a business and needs to produce as much stone as possible to generate profits. It is up to the stonemason to make sure that the best possible deal is achieved when purchasing stone – this means minimizing your time getting it ready for working. So when taking possession of stone from the sawyer check for the following:

- Saw-lash. This occurs when the blade of the saw moves in and out, leaving ridges on the face of the stone; this can be exacerbated if the cut is made too quickly or deep in a hard stone.

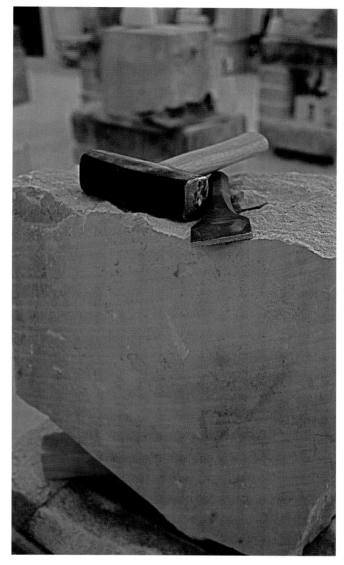

Block with sawlash.

- **Arrises.** If the arrises are not protected they can become damaged very easily during handling in the yard and loading into transport. When the stone needs a sharp arris make sure it has got one. Ensure adequate softening to protect the stone when it is stacked and loaded.
- **Orientation (bedding)** of the block, which should always be specified. This can be checked visually with many stones, but with some you will have to take it on trust.
- Get used to checking stone by tapping it with a metal bar (chisel is good) and listen for a dull report that can indicate vent or a fracture. Stone that rings true will be the one to use, so start listening to what the stone tells you.
- Finally, make sure you are able to get the stone home, or that you have factored in the cost of transport.

Industrial size stonemasonry workshop – far removed from a cottage industry.

- **Stepped face.** This can be cut accidently, which gives two levels on the same face. Unless any of this is on the back of a stone and doesn't need to be scribed up, the first task will be to get rid of these marks by working down before working can commence. A good sawyer will regulate the speed and depth of cut to allow for the stone, ensuring a smooth finish.
- **Dimensions.** Check that they are as ordered. If they are cut too big, you will have to work them down to size. Make sure all faces are square to each other; otherwise templets cannot be relied on – so there could be even more working before starting.

The best way to move stone.

CHAPTER 2

WORKSHOP AND TOOLS

The stonecutting work covered in this book could probably be carried out on a sturdy table in the kitchen, mess notwithstanding. In practice, however, if you are here to learn and carry out work in a way that makes this a worthwhile experience, it is best to do it right from the start. This requires a setting known as The Workshop. In a perfect world, this is how it should be…

The workshop is, in all probability, the most important place that an artisan will have a hand in designing, for out of the stonemasonry workshop can come much that contributes to the built heritage of our world, as well as providing a living income. As it will certainly be the most important place in your working life you must contrive to make it an area that gladdens the heart when you enter and encourages work without stress. A workshop in a neglected and untidy corner of a shed is a deplorable state of affairs that could only reflect negatively on the competence of the stonemason and the quality of work produced there.

Apart from time spent sleeping, the stonemason will probably spend more time in the workshop than any other place, and as such it should be considered in terms of comfort and ambience, rather than the scene of penance that seems to be the way of many. Consider that from mess and confusion, beauty and clarity seldom arise, and if they do it's plain hard work.

safe manner, which can only be achieved if the working environment is conducive to safety, productivity and wellbeing.

As a (possible) entrepreneur you may need to employ help and you also may have visitors in the workshop or on site. It therefore becomes your responsibility (in law) to ensure that anybody within your realm is kept safe from harm, principally accomplished by providing a suitable environment, detailed instruction in the use of the workshop and if necessary training for equipment.

Well-illuminated and airy rooms, though sometimes difficult to attain, are desirable (the reason for this will become obvious once work is started). Ample natural light is to be preferred over artificial, but unfortunately this is often impractical. So aim to install lighting that is conducive to the ambience – remember you may be looking at a pale piece of stone for days on end, so no flickering neon or tinted low wattage bulbs.

Headroom must be taken as it comes, but if opportunity allows make the height substantial. This allows a lot more work to be carried on before it gets too dusty, and it could allow the installation of useful lifting gear. It is commonly held that airy, well-proportioned rooms contribute favourably to personal wellbeing.

The Portland workshop.

WORKING ENVIRONMENT

As an artisan you will have a duty of care for your personal welfare and a sensible desire to carry out your livelihood in a

OPPOSITE: Using a disc cutter to remove waste stone prior to working; power tools are now an everyday part of the modern stonemason's life, but it is still the hand skills that will create the finished piece.

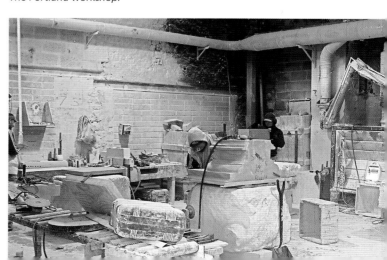

The floor should be smooth, level and hard, preferably concrete or well made with solid board/timber planking sturdy enough for stone that will need to be moved about on wheels or rollers. The health consideration is that standing on uneven surfaces for long periods will cause physical discomfort and health problems. The accumulation of dust and debris will need regular cleaning, usually with a shovel (it will destroy standard vacuum cleaners), so if the quality of the floor is poor to start with, it will get worse – this needs to be the best you can achieve.

Once the floor is suitable for the task of surviving in a stone workshop, you may experience – through standing for hours on this hard, unyielding and possibly cold surface – discomfort that will affect rate of work. The simplest way to check this is to stand on a cushioned surface such as carpet or a rubber mat. This is now a recognized issue, and specifically designed mats are readily available for working all day in a standing position, but as a stonemason is always in possession of squares of carpet one of these will do the trick.

ATMOSPHERE AND LIGHT

Having good ventilation is commonsense, as long it does not impede the work by causing winds to stir up dust or body chilling draughts.

If climate allows, a workspace set up outside in a covered area can be practical, and will be healthy in terms of ventilation and light – though try not to work near vehicles, as

Make-do banker – plenty of clearing up to do.

the abrasive quality of stonedust will quickly destroy polished paintwork.

Dangers of Inhaling Dust

Be aware of the nature of the stone being worked, as it may be necessary to get some protection from inhaling harmful mineral content – specifically silica from sandstones. For any stonemason who is to be working regularly on stones that can be deleterious to health it is important to employ a suitable means of dealing with this. Personal protection ranges from dust masks of varying efficacy to all-encompassing pressure-fed helmets where the forced air pushes away the dust and provides a clean atmosphere to work and breathe in.

Obviously the simplest method would be to prevent inhalation of this material by using a dust mask but, being pragmatic, working for hours or days wearing a protective mask is not a pleasant situation (or practical as they become less effective each time they are removed and replaced in the course of the day). In view of the prospect of contracting some debilitating and possibly fatal lung disease, some research on the most appropriate methods available is vital.

In the best scenario, an industrial quality dust extraction system would be installed. The costs and maintenance will make this viable only for those that can afford it, although portable extraction units can be hired. Consider that cost should be secondary to whether the system can do the job properly – so do some homework and prolong your life.

Good quality dust mask with interchangeable filters. Choose specialist filters for working with resins or solvents.

A reasonable stopgap, and probably the method most often used by healthy, fresh air loving artisans is to get the stone outside, preferably on a bracing day, and (sporting a comfortable dust mask) get stuck in.

EYE SAFETY

While irritation of the eyes by airborne dust is another complaint for workers in this industry, far more annoying – or hazardous – is getting a fragment of stone hitting your eyeball at speed every time you strike the chisel. It is advisable therefore to give serious consideration to suitable eye protection, the emphasis being on suitable in the absolute best possible sense, as eyes are not replaceable commodities.

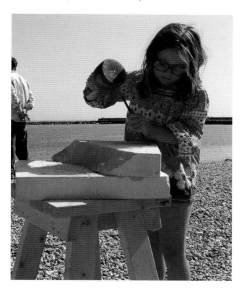

Blossom (the author's daughter) working on the beach.

Attempting to work inside a closed room wearing a dust mask and ill-fitting cheap glasses will soon become a pointless task for one or more of the following reasons. The glasses will fog up as the dust mask restricts the flow of cool air, creating a small localized mist that condenses inside the lens; wiping them will smear and obscure them; this sticky surface will become coated in dust that will abrade the lens and can completely ruin any transparency in a day. The glasses won't stop shards of stone flying into the eye – consider that this may only happen once or twice – but you will spend the rest of your day blinking every time the stone is cut or gaining a continuous squint that in the short term leads to headaches, and worry/frown lines for the future.

Personal preference will obviously come into play here, but paying out for a top of the range pair of protective glasses will be cheaper in the long run than continual replacement of much cheaper items, with the added benefits of prolonging eyesight and knowing that the only lines that will gently mar the corners of your eyes will come from laughter.

Lenses made of glass rather than plastic will give longer life

and remain clear, as the stonedust will not haze the surface so readily.

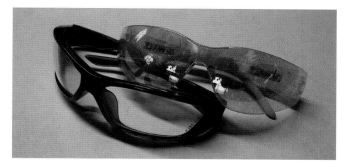

Safety glasses. Note that the stylish ones are all scratched up from stone dust (and quite pricey as well). Always choose good quality lens material and clean them off each night by rinsing or blowing with airline.

Another aspect of eye protection is the quality of light. During daylight hours working on stone outside is a good solution, especially if it is sunny or bright. If the stone is pale then it is noticeable that a mild form of 'snowblindness' can occur, giving rise to headaches and eyestrain. In this situation a pair of tinted glasses should be worn. Remember though, they still need to be of safety standard.

Johanna (carver of Spirit of Portland) working in Bowyers Quarry, Portland.

Protecting the Feet

A very important item is sensible footwear, which should always be worn whenever at work, as anything that falls down could land on your foot at some point, and there are a lot of things waiting to prove the attraction of weighty things for the down direction – from relatively lightweight chisels to the blocks of stone.

Don't be fooled into thinking you can cope with the occasional act of gravity by just yelling and applying a plaster – things can get really serious. (Before the use of safety footwear became universal my grandfather dropped a brick on his foot, causing a fatal blocked artery from the resulting complications. This was the particular incident for British Gas to adopt steel toe-capped boots.

THE BANKER

The banker (from the French *banc* meaning bench) is the hub of production and needs to be centrally situated in the workspace, allowing access from all sides – it is easier to work around a fixed stone rather than keep repositioning it to get at the side being worked on.

Construction of a sturdy nature is essential – it has to take a lot of weight. A fixed banker can be made from concrete blocks, either cemented together to form a solid mass of appropriate height or dowelled together using steel bar with pieces of carpet laid in the joints to absorb vibration. A solid cast block that can be lifted up and moved around when necessary is good, but a better option is to have one you can alter. Consider that there will be much pounding and vibration, so ensure that whichever construction method you use can take the punishment.

Portable bankers can also be made to many designs from timber, the criteria being strength of joints and stability when stone is struck. It is useful for them to be of a size and weight that can be transported to site for lifting onto scaffold, and also weatherproof.

Banker with dowels and carpet padding.

Portable bankers.

Solid banker.

The approximate dimensions of a useful banker would be level with the waist in height, with a working area of about two square feet and a shelf underneath for tools. On a wooden banker a bracing rail around the base will give a rest for one foot whilst working – leaning into the work with an offset foot stance (like upright surfing) will be balanced, often more comfortable than standing 'duck footed'.

For obvious reasons, now you have designed your solid bench, the height of the work surface would best be adjustable. This can be achieved by adding height to the banker using sandbags, concrete blocks, slabs of stone or timber. Never attempt to work from a higher level by standing on a loose construction of blocks as in the inevitable event that you stand back to check (admire) the work, accidents will happen. It is also uncomfortable to work from a shifting floor. So the banker is best low and built up rather than high and unattainable without a platform.

Alternatively use a 'hop-up' – a sturdy platform purpose-made to stand on and work from. These, made in a variety of materials and styles, should always be robust and large enough to be safe.

Whatever the work surface of the banker it will need some form of softening to protect the stone edges and prevent the block sliding around. Sample carpet squares, easily available from a furnishing shop, are the most economical and practical method to cover the work surface – though they do collect dust.

HOLDING SMALL THINGS

If the stone is small it may be difficult to keep it in one place whilst working on it. This can be resolved by filling a washing-up bowl with sand and resting the stone in it. This is a common aid for carrying out small carvings that need to be shifted regularly or do not have a flat surface to rest on. Sandbags will perform all these functions as well (though you cannot use them for mortar afterwards). Inner tubes filled with sand are a good alternative.

TECHNIQUES FOR MOVING STONE

Moving a block of stone, from outside the workshop, onto the banker can be either a trial of strength or a simple application of mechanics; it is personal choice whether to strain intelligence or muscles. The best advice is to be one of the few workers in this industry who endeavour to live an active life, unhampered by backache and crushed digits; get thinking – let physics and technology do the work. Health and safety information on correct posture and procedure for lifting is widely available, so make use of it.

First be aware of personal limits compared to the weight of stone. Usually any block of stone larger than a basketball will be too much for the average person to dead lift from the floor, so if it is possible to lift with a colleague do so; alternatively figure out a way that will not hurt you.

Once you have lifted the stone manually or by hoist, you

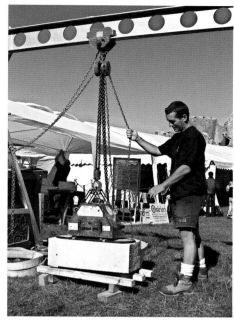

Stone 'magnet' works well on stones with a fairly even surface.

John Ashurst, one of the most influential people in the restoration of historic buildings, here demonstrating his superhuman prowess when moving stone (This was actually a fake column at a lecture.)

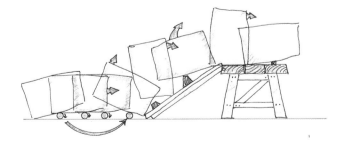

Progress of a block across the floor on rollers and then up a ramp to the banker. Never lift the stone straight up; use bodyweight to tip it onto corners (remembering to put softening down first). Slip a roller underneath and push to the point of balance. A second roller is placed under the leading end and the stone rolled onto a third roller. Once first roller is freed, move it to the front and repeat the process to get the stone across the workshop.

Pump-up trolley for moving stone; large and small disc-cutters for stone.

Portable gantry out on site.

Shifting blocks with HiAb and slings.

must be able to put it down correctly allowing hands and straps to be pulled out in one piece. To do this, quarter the stone when placing it on a banker by resting halfway along its length tilted on the edge of the banker so fingers can be removed, then tipped flat and slid into position. If the shape or size makes this awkward a variation is to place a roller on the banker near the end of the stone and use this to get fingers clear before rolling the stone off.

Loading the stone direct from transport onto the banker is always preferable; using sturdy planks and rollers involves no overly excessive exertion. This does rely on being able to get the vehicle into the vicinity of the banker and on heights not being too different. Trolleys with flat beds can be put to good use here, as long as they can move around on all the surfaces; so wheel size and design is important. If the platform can be raised and lowered it is even better.

Portable gantries can be useful investments if large stones are to be handled regularly and will be useful on site for lifting slabs and blocks into place. For large workshops or if there is suitable transport, even better is a hydraulic or mechanical crane – this is getting away from the small workshop but not unfeasible. Consider: if some large stones need to be moved, perhaps from bankers to transport for offloading at site, the cost of hiring machinery to allow one or two people to accomplish this quickly without strain may compare favorably to employing extra hands, or taking time out from actual working through everyone convalescing with bad backs the next day.

TOOLS

In this age of cheap manufacturing processes, and ubiquity of poor quality tools, it is satisfying that the relatively small demand for essential tools used by stonemasons means they are, by and large, produced by specialists concerned with quality and not quantity. So even though they are not cheap, you can – and should – aim to become the owner of good quality tools that will survive your working life as a stonemason.

Buying cheap tools is false economy, for compared to other crafts they will be employed in a harsh environment, and will need to be fit for purpose at all times. The following recommendations are based on your having purchased the best quality implements.

STONE MOVING CHECKLIST

As every workshop and the stone to be worked will be singular it is more relevant to present the important points rather than trying to describe a specific situation.

Good Practice

- Ensure a clear passage for moving stone; remove trip hazards and loose materials.
- Wear protective toe-capped footwear.
- Wear gloves to prevent a slipping stone tearing skin.
- Use trolleys and protect all surfaces of stone from bare metal.
- Have a supply of wedges or blocks to hand, to rest stone on and aid the removal of straps and fingers.
- Use correctly designed and appropriately strong lifting straps.
- Ensure all lifting mechanism is correctly serviced and functioning.
- Always store or rest stone on timber beams that allow you to get a purchase underneath; use softening, as stone has very fragile arrises and hard wood can damage soft stone.
- Keep the bench as clear as possible, tidy up tools.
- Protect the edges and clean surfaces of stone from damage – especially when the masonry has been finished and is on its way to the client.

Bad Practice

- Attempting to lift too large a piece of stone without help.
- Using a wheelbarrow for transporting stone.
- Using crowbars or other metal levers without protecting the contact points to prevent damage to the stone.
- Dropping stone onto floor.
- Stacking stone on unstable supports.
- Forgetting to clean up the debris from working each night – surprisingly small lumps of stone underfoot can throw an early morning worker off balance.

Not the tidiest of tool arrays.

Remember: The design and running of the workshop has as much to contribute to the production of quality work as the tools and skills of the stonemason.

The vast range of chisels that can be employed by the stonemason.

Secondhand tools are often shunned by callow, inexperienced workers with an eye towards the shiny and new, but good quality stonemasonry tools – especially mallets, hammers and chisels – can be expected to outlast their owner. So keep an eye out for the chance to buy some secondhand when they come up; you might get tools with provenance and possibly meet their previous owner; if this is the case – lucky you - have a chat and get some tips!

In the buying process look for how the tools have been maintained. Chipped chisels can have the blemishes sharpened out or reshaped (as long as there is enough metal); hammers will accept new handles; and disfiguring surface rust can be removed with wire wool and light oil. Don't dismiss rooting about in buckets of seemingly scrap tool bins – the occasional gem can often be found, to be revitalized with some elbow grease.

Three areas of work undertaken by the stonemason need dedicated tools – setting out, working of stone, and fixing – though there is often a requirement for some to be used across the board.

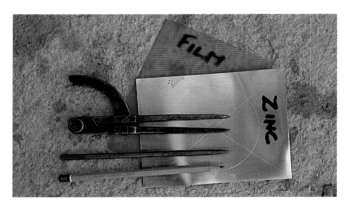

Scriber, dividers, zinc and film.

Setting Out and Measuring

Drawing instruments made from a durable material like steel are preferable to drawing office plastic which, unless it is kept away from the workshop, will become scratched and unreadable very quickly, though admittedly the abrasive qualities of stone dust will also dull up metal surfaces.

Metal straight edges and squares can be used both in the working of the stone and in the setting out; just ensure they are cleaned and dry before using them on paper.

Pencils used for drawing on paper should be grades of appropriate hardness for engineering drawings, pencils used

SCRIBERS
Scribers are an essential item in the drawing office, workshop and on site, but due to size (and ease of borrowing!), they are prone to loss. A good substitute for these relatively expensive steel pencils is to get hold of an old chainsaw file (once worn out they are readily discarded) and on the tungsten wheel of the grinder give it a sharp point; the metal is super hard, and will stand years of sharpening, while the handle gives a location for your moniker and stops it being lost.

on stone will require a hardness of H9/H10; these are too hard for paper and should be kept separate.

Curves
You need compasses with a pencil or pen holder, and dividers of sturdy construction that can be locked in position with sharpened hard points. For larger arcs, beam compasses or specific heads can be bought to use with lengths of wood or metal bar.

Developing sections of mouldings will require arcs that need to be accurately drawn freehand (which takes practice) or by French curves; these come as a set of three templates with a variety of developed arcs and curves, usually plastic – so look after them.

Cutting
Templets and moulds for setting out on stone are traditionally zinc or now plastic film.

If zinc moulds are to be the preferred style then curved and straight tin snips, fine metalworking files and wire-wool will be essential to the kit box.

Plastic film templets can be cut out with good quality scissors and sharp hobby knives (replaceable blade type). These can be stored and transported rolled up in cardboard tubes (such as from cooking foil) or short lengths of plastic tube – white is best as the description or reference number can be written on the tube with a permanent marker (another essential stocked in varying nib thickness) which saves having to get all the templets out when looking for an individual one.

Snips for cutting zinc.

Set of soft stone chisels with brass hammer. These are not woodworking tools – they have metal rings to prevent the handle splitting.

Working of Stone

Chisels

Chisels for stone are initially described by the type of head: mallet, hammer headed or turned shanks designed for pneumatic air hammers.

Secondly chisels have one of two types of blade material for the cutting edge. Traditional fire sharpened chisels use an edge that is drawn out of the (steel) shaft in a forge where the blade is heated and quenched to give a tempered metal ready to be sharpened. The other material, tungsten carbide (TCT) is a very hard metal used in more modern chisels. Its ability to hold an edge better makes it the preferred choice of most, and it is also better for more abrasive stones. A TCT chisel has a piece of the metal brazed onto or set in the end before being shaped and sharpened.

Medium chisels. The top (mallet head) and bottom (hammer head) are fire-sharpened; the centre is TCT with a removable cap for use with a mallet or air-hammer.

Wooden handled chisels similar to carpentry ones, with extra features to cope with the pounding they get, are used extensively by stonemasons in regions of softer stone. These are not carpentry tools so do not attempt to use woodworking chisels on any stone, as they will soon be destroyed and unfit for use in either field.

Finally chisels are named by the shape and size of the blade. There are chisels available for every application (straight bladed chisels wider than 1½in/40mm are termed boasters. To buy one of each would be crippling in financial terms as well as being impractical in storage and lugging about on site; the exercises in the following chapters can all be undertaken with the list below. So consider these fundamental to start off with, then build up a collection. As an artisan you will come to appreciate and covet tools of quality. You will also no doubt end up with some chisels that will never be used, though they will look good and organizing tools is a pleasant pastime.

Drill, for use with brace or hand.

Mallets and Hammers

While traditional mallets are wood, there is a tendency now for them to be of hard nylon/plastic derivatives, varying in size relevant to the work. Wooden mallets tend to wear out, fracture or disintegrate, whilst the plastic ones remain stable for a very long time, are easier to source and can be tailored to individual needs by turning off material.

Wooden mallets. These are made from lignum vitae – a very hard wood; the smaller was turned from a bowling ball.

Modern mallets of nylon/plastic. These are better than traditional wood as they do not have to be used in specific directions.

Pitching or mashing hammer. This is the type to use with a long head to provide a lot of force without bulk.

The correct wooden head for a mallet will be turned from a block of hard wood that has the grain running from one side to the other (apple, cherry or box are some types used). Do not buy one with concentric rings, as although this would seem the logical way the design is essentially useless. The chisel is only hit with the end of the grain, which limits the striking area to opposite sides of the head. Striking along the flat sides will cause it to degrade quickly as the wood curls round the strike point and layers off.

So when buying secondhand see how the wood is holding out. Remember that this is a tool to be used, so don't buy a mouldy old bit of kit that has reached the end of its working life no matter how authentically interesting it looks. Unless, like me, you want it as an ornament in the workshop.

A hammer, basically a block of metal of various weights on a wooden shaft, gives the stonemason a smaller, weightier tool than the mallet, with a very different working performance. Shapes vary from Fischer style where the head concentrates the weight of the metal to a small area, lump hammers with a square or rectangular head, to a dummy which has a turned cone of metal traditionally used for fine carving and lettercutting.

Often some stonemasons will prefer to use a hammer rather than a mallet for all work. All these tools have developed over the millennia for working in specific situations, with particular stone or to carry out types of work, so to dismiss a particular type is foolishness and more stick-in-the-mud than skilful.

Axing is another method of working stone, which uses such as a tallion (French: tallion de Pierre = cutter of stone) or stone-axe, probably the earliest specifically made stonecutting tool. The marrying of a heavy stone or metal head to a wooden shaft draws a line from the origin of all toolmaking. Unsurprisingly axes are still used extensively and effectively across the world.

The axe combines the qualities of hammer and chisel, allowing rapid removal of stone. In the right hands axing can be incredibly efficient and surprisingly accurate for producing quite complex shapes. Axing is now almost a separate

Stone axes.

skill from the stonemasonry taught today, but, if the chance arises, it is quite useful to understand the technique as obviously many stones used in historic building would have axed finishes. These can be replicated by deft chisel work, though much more satisfying when worked in this atavistic manner.

Ollie giving a demonstration of axing off a flat surface.

Working column with air tools; notice the protection needed and anti-vibration gloves. Modern practice is not to use this type of tool set-up for long periods, in order to prevent white-finger.

DRY SHAFT

Wooden shafts can dry out if kept in an air-conditioned shop, and become loose in the head as they shrink. For a quick fix just place it head down in a bucket of water and the wood will swell to become tight again – you may have to do this occasionally in the workshop or on site. Always keep the shaft locked in tightly, by using correct wedges and checking for breaks in the length. A kilo or more of metal flying across a workshop is a dangerous situation.

Pneumatic or air-hammers are widely used in stonemasonry, they require a source of compressed air and specialized elements, the purchase and installation of which can be financially substantial. There are also health implications to be considered, as the noise and dust generated is not pleasant, whilst continued vibration through the hammer can cause a serious, debilitating medical condition known as Hand Arm Vibration Syndrome or commonly 'White Finger'.

Straight Edges, Drags and Files

Straight edges of various lengths, essential for the monitoring of progress and accuracy, should be of aluminium, hardened or stainless steel. They should have a bevelled edge and a hole at one end for hanging up – left on the floor or leant against a wall they can be easily bent enough to be useless. Metal squares of whatever sizes necessary should be kept in a protected area, as dropping them can put them out of true.

Do not use any other material than metal to make a straight edge, as the constant abrasion by sliding over the surface of the stone will soon render them inaccurate; aluminium will be affected like this so monitor the accuracy.

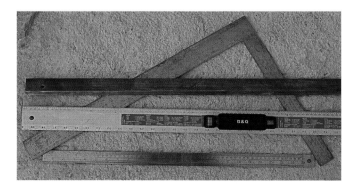

Straight edges and square.

TRUE AND SQUARE

Before starting a new job, or if there are concerns of accuracy, check how true straight edges and squares are.

Place the straightedge on a flat clean surface and, with a sharp pencil, make a line along its length. Then rotate it 180 degrees and draw over the line – these should match up exactly. Also, start the same way, then turn it over and draw over the line to check.

Place the square against the straight edge of a board and draw a line along the inside of the blade (the outer edge is not used for marking out) on the board, turn it over and place the corner at the start of the first line and draw another along the blade; these should match up exactly.

The cost of machining crooked straightedges and squares is excessive against the cost of replacement, so if they are bad get new ones.

Flexible diamond pads mounted on dense foam blocks. These are a good investment, as they should never wear out if looked after, and can be had in a wide range of grades.

Drags and files for stone. Flat bladed drags are good for dressing in stone when fixed. The handled (French) drags are for removing large amounts of soft stone rapidly.

Sinking square and sliding bevel will handle aspects of monitoring depths of fillets and accuracy of chamfers between them, as well as transferring setting out information to the stone itself.

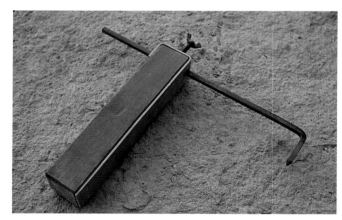

Box trammel. The brass strip prevents the sanding away of the wood against the stone.

Drags, files and rubbing blocks of various grades should be used where the stone allows. Hard stone drags often have TCT cutting edges, to aid in levelling, cleaning and finishing; generally these are used after the chiselling has been completed and can impart good texture to softer stones.

With softer stones, French Drags (*Chemin de fer*) are often used to remove waste stone efficiently – the hideous racket these produce will not be popular with anyone in earshot, and use should be limited appropriately!

In the corner of every workshop should be fixed a well-maintained, sturdy bench grinder with wheels suited to steel and tungsten, for shaping tools, for chisels prior to sharpening and honing on whetstone or diamond sharpening block. Mount a cup holder next to it for quenching and cooling hot metal.

BASIC TOOL SET

For working stone you will need at least one of each tool group to get started. The photographs show a variety and there are many variations of detailing, but if it is of the basic design the tool will do the job.

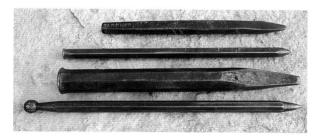

Punches and points. The second (point) from the top is TCT.

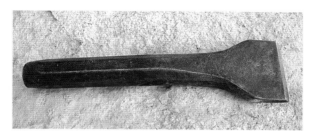

Pitcher or hand-set. The blade is only chamfered on one side to provide a heavy controlled strike.

Inch and half-inch chisels.

Claws.

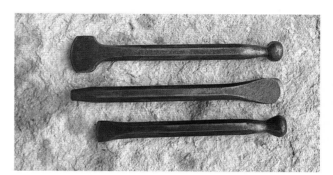

Bullnoses. Top is TCT.

Boasters. Top is a mallet head, fire-sharpened (over 125 years old and still in use), centre can be used with mallet or air-hammer.

LEFT: *Square, sliding bevel and sinking square.*

Saws

On some stones hard point saws for general carpentry will be good enough; with a thin blade they can produce very fine joints.

For the harder stones tungsten teeth are used on dedicated saws – indeed drags and French drags can all be purchased with these tips. The saw will cut quite a wide channel and needs to be handled with control to get best results.

Tungsten teeth saws.

Fixing and Site Tools

As we enter into the comprehensively supplied realm of building site equipment it would be easy to note what a local builder has in his toolbox and replicate it – but no such luck! The fixing of stone into a building, especially one that is already made, needs to be conducted with the right tools and intelligence to produce harmonious work, neither of which tend to figure high in the average construction worker's arsenal.

Spirit levels and straight edges are obvious for getting work to line up. A chalk line and plumb bob will both need to be used – look after and don't bend them to keep them accurate. To prevent a wasted hour of untangling always store string lines securely wrapped round a bobbin and secured by a rubber band; five minutes doing this properly will allow you to earn more money.

Trowels

The range of shapes and sizes of trowel can be bewildering; best to have a range of these. Small leaf or spatula trowels used by stuccoists (ornamental plasterers) are for fine pointing. A small diamond bladed trowel used by bricklayers should have the sharp end rounded off for coarser pointing and manipulating mortar. Gauging or brick trowels are needed, and bucket trowels for mixing and placing mortar. The skinny

bladed trowels for fine pointing can be useful in ashlar work. With all these, experience, need and preference will dictate the final choice.

BEST TROWEL

When choosing a trowel always buy those where the blade and tang are forged together, that is made in one piece. A tool destined for a hard life made thus is far stronger and preferable to welded items (the weld is strong but creates weakness around itself). Inspect the tool and if you can see any pinpricks indicating poor quality metalwork leave it be as it will invariably rust out.

General trowel and fine pointing trowels, which are commonly known as plasterers' tools.

Hawks and Wedges

Using an appropriately sized hawk for holding mortar while pointing or repairs are carried out, is preferable to ungainly balancing the mix on an upturned trowel and trying to slide it off with another. A hawk can be simply made from wood, purchased as a GRP one piece or a wooden handle with a metal plate.

A small collection of old saw-blades from hacksaws, tenon and wood saws are great for cleaning out joints of mortar and dirt (more of which later).

Wedges of various sizes will be needed in many situations and should be designed as folding pairs. Basically take a rectangle of wood and cut diagonally to produce two identical wedges that have the advantage that, when placed longside together and moved in, will lift perfectly vertical. Made from hardwood, sanded smooth and possibly waxed, they will slide easily and are waterproof to boot. Make them as needed, or when you come across a suitable piece of wood; mark and keep them in pairs.

Small mortar hawk.

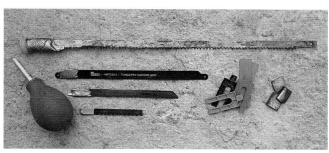

Various saw-blades for cleaning out joints – one end has tap and foam over to allow their use without cuts and scratches; blower to clean out dowel holes; plastic and lead packers.

Hardwood folding wedge. Keep these together in marked pairs (and do not use them for general wedging jobs!).

TOOLKIT

The tools mentioned are what you need to undertake all the functions of a stonemason from design, production and fixing. From the off you will begin to accumulate more tools almost every time you are near a shop or market that sells decent utensils.

Armed with the above you are now properly equipped, so it makes sense to look after them by getting a sturdy toolbox to carry all the hand tools about and keep them secure when not in use. Buy for sturdiness (to sit on, rest stone or as a hop-up) and simplicity rather than plastic gimmicks, as the life of the stonemason's toolbox is a hard one.

Toolbox, sturdily constructed with braced corners. It can perform a lot of functions on site; the hasp for a padlock is a wise extra.

For moving about on many surfaces and hoisting onto scaffold, solidly attached lifting handles are more useful than wheels. A lockable lid is better than multitudes of drawers and trays ready to spill out when the mood takes them.

All metal tools should be clean and dry when put away. Wiping over with a light oil-soaked rag is sensible practice, but remember to dry wipe them before using again, as oil will stain stone.

Have a tool roll for storing chisels and small tools, to keep edges keen, stop them rattling around and help in picking them out. If left loose, the bottom of the toolbox will quickly collect a layer of dust that needs to be emptied out regularly. The best material is canvas, as dust can be removed easily.

Straight edges and levels should be stored out of harm's way by hanging up in the workshop. When used on site a section of plastic gutter pipe with a cap on makes transporting them easy and keeps them in one place.

KEEPING HOLD OF THEM

By now a considerable sum of money will have been spent on some very desirable items. Unfortunately they may be desirable to others as well.

As you will probably be involved in working with other people on sites where the opportunities for 'borrowing 'and worse 'forgetting to return' exist, you (and the 'borrower') must be able to identify your own possessions easily without resorting to micro-examination or argument.

Permanent markers only earn the title on smooth surfaces that are not handled frequently or areas that cannot be wiped clean with any solvent to hand on a busy site.

Use paint on items such as toolboxes, where the removal will be difficult or at least leave an incriminating mark if attempted. Otherwise hand tools need to be marked by impressing or engraving; before starting this oddly satisfying task choose a totem that can be readily identified and applied to all the tools in your box with equipment you possess – so keep it simple. A box of letter punches is cheap, and initials can be stamped on appropriate spaces – failing this, use a drill or dremel to cut in a design – once again keep it simple.

It is wise to insure tools against loss (claiming for damage is a difficult one for hand tools) in the unfortunate circumstances of theft or misplacement. To alleviate the awkwardness in proving ownership it is handy to have all your equipment listed and photographed, with close-ups of unusual details as well.

[Note: Items considered consumables for day-to-day projects will be mentioned or described in the pertinent chapters.]

SHARPENING AND GETTING THE EDGE ON CHISELS

Watch any stonemason pick up a chisel for inspection. It will be gripped loosely in the fist with the blade up by the thumb, then the thumb will be run along the edge – voila the inspection is complete! For what counts here is that they should

have encountered an edge keen enough that with minimum pressure could draw blood. If that need is not met then the chisel needs work before it can be used for anything more than a wedge or lid lifter.

The edge of a chisel should be straight and with an even bevel along its blade. A misshapen edge will never make a flat surface, so sharpening is to be undertaken with the usual remits of control and accuracy – all done by eye.

Fire Sharpened Chisels

Chisels should be started off on the steel wheel of the grinder: stand comfortably, support your wrist on the rest and holding the chisel firmly, at the angle desired, run it squarely once across the rotating wheel watching the edge: a stream of

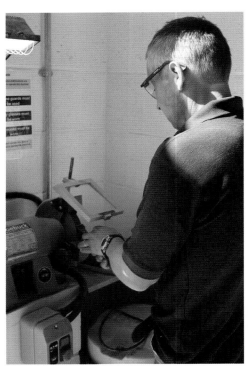

The first grind.

sparks (eye protection!) will show the abrasion happening, and by keeping this constant will result in an even chamfer across the blade. Rotate the chisel and do the other side. Don't go too fast as the chisel will bounce and the grinder will score the edge – working steadily will get a better finish.

The durability of the cutting edge relies on the tempering received in the forge, which will be lost if excessive heat is generated by grinding. So keep a container of water in which

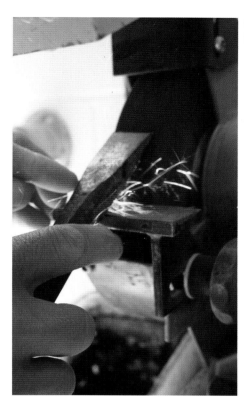

Correct angle for grinding edge.

Honing boaster. Note the angle of the blade.

to regularly dip the chisel during grinding; this is an essential practice so make the container fixed and always replenished – a sloppily placed plastic cup will not stay around long, especially in the vibration zone of the grinder.

When satisfied with the result, move onto the sharpening stone, generally nowadays a diamond impregnated pad. Grip the chisel in the fist (similar to holding it for work) to keep it at the right angle, then using the other hand to push as well as applying constant downward pressure, slide the chisel forward along the stone in smooth sweeps. Repeat this a couple

The angle for honing the edge on a diamond pad.

of times and inspect the result: the chamfer should be of an even width, dead flat and of an even non-reflective tone (this will take longer to achieve if the primary grinding is carried out too coarsely).

When the other side is finished, sight down the cutting edge to check for perfection and repeat till the blessed state is attained. Everybody will tend, in the beginning, to sharpen with a bias to one side, which will result in a narrowing of the chamfer; this can be corrected by twisting the wrist to apply extra pressure to the narrower end whilst sharpening.

Tungsten Carbide

TCT edges are to be ground in on the appropriate wheel (usually coloured green) of the grinder in the following manner. Assume the same stance and grip and slowly move the edge along the wheel – there will not be flying sparks like the fire-sharpened, just a small starburst of colour at the edge to indicate where the action is taking place, so keep this constant as before.

The material is much harder and will not grind so easily as steel – do not force it as this will result in uneven wear. The structure of tungsten carbide is such that rapid changes in temperature will cause a loss of matrix resulting in the edge crumbling away, so getting it hot and then plunging into the water is not a good idea. After each pass on the wheel place

TCT on green wheel – no flying sparks here.

Bullnose sharpening. Start on one side...

Rolling the chisel.

... and rotate to the other.

the edge into your palm to check the temperature; if it is too hot to bear, stop grinding until it cools down – never quench it.

Hone as fire sharpened.

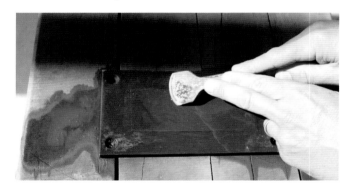

Bull-nose

The other chisel that will be used in this book is the bull-nose, which carries a blade that is curved across the cutting edge (these are useful in the shaping of stone, particularly in architectural carving).

The bull-nose requires a singular sharpening technique. The primary grinding on the wheel is by holding the blade to the wheel and rotating the chisel around the shaft. The position of the chisel stays the same, and by this method the blade can be ground in a controlled manner; be aware that sharpening

These three images show stages in bullnose sharpening. The blade descibes a figure 8 in order to hone evenly.

Do not move the chisel shaft in a pendulum manner to rotate the blade on the grinder or block – it might seem logical but will never give the results desired.

aggressively can change the shape of the blade or with TCT reduce the amount of cutting material.

On the sharpening block hold as before and describe an 's' while pushing forward along the length of the block, rotating the chisel so that all the edge comes in contact at some point during the process and is sharpened.

The Sharpening Stone

After time the abrasion from sharpening will start to wear the stone unevenly and this will in turn make sharpening chisels difficult, so it is essential to maintain a dead flat surface on sharpening stones, whether they are oil or water stone. They can be ground flat by rubbing against a similar or harder stone (ex-grinding wheel discs are good for this, and useful for rubbing down masonry work) or if the wear is too much get your local sawyer to cut a new face on it.

Or simply buy a diamond sharpening pad of a quality that will last years. Mine is twelve years old and shows no sign of wear and will last for many years to come. Dividing the cost by lifetime will show a saving over buying a cheaper sharpening stone – and it's a lot easier to transport, use and look after.

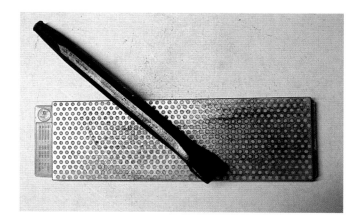

Quality diamond sharpening/honing pad.

SETTING OUT AND MOULD MAKING

Planning and preparation are crucial steps to getting things right. There are few situations in life where this is not necessary, but in the workshop it is obligatory.

In the production of a building or monument the first step for the stonemason is the setting out, in which the design is divided out into convenient parts and then all the relevant information is produced as a templet or mould and then subsequently transferred to the stone.

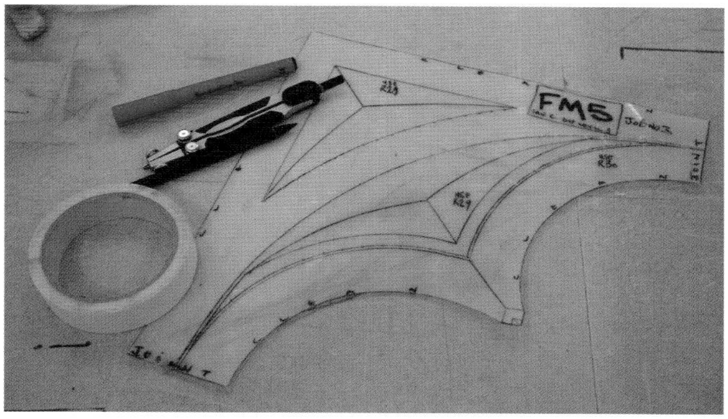

Templet for quatrefoil.

Producing a piece of stone that will be fit for purpose requires precision and thoroughness in the planning stage. As an assembly of single components a building or monument must have stones that fit together with the minimum of fettling. So a simple error made at the setting out stage will generate more work for the stonemason, whilst imprecise stones could have a detrimental effect on the whole.

OPPOSITE: *A church built of local Devon sandstone. The main structure is what can be described as vernacular. The simple design of the doorway and the top window are strong examples of early medieval geometry, which contrasts with the stiff formal work on the much more recent window; work that reflects the experienced stonemason of the period and his skills in setting out.*

EQUIPMENT

The Block

The stone for a masonry piece, if you are not shaping your own, is usually purchased as a block sawn six sides (SSS) from the stoneyard, an exact fit to the outer dimensions or extreme points of the design onto which the templets and moulds can be applied. It is down to the sawyer's skill and experience to get the most economical cut from the stone that will ensure minimum waste after working.

Checking for square – always work from one side as a reference.

Scaled Drawing

To begin with, a geometrically correct and scaled drawing is made of the stone in orthographic projection; this will show the face (front elevation), top bed (plan) and section through (end elevation).

From this drawing the full-sized templets or moulds are produced. Traditionally made in zinc sheet, nowadays stiff plastic film is more practical to use – both materials are stable and unaffected by moisture.

If the templet is cut from porous materials such as paper, cardboard or hardboard, they should be used immediately as the absorption of moisture can cause them to become damaged, distorted or change size, which will mean that subsequent stones worked to them can be incorrect. So it is not recommended but occasionally needs must.

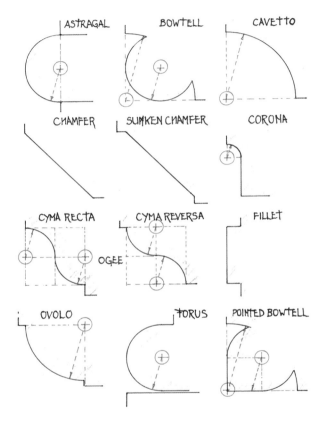

Common mouldings.

The Setting-out Table

A large table dedicated purely to setting out would be a luxury in a small workshop, as you will not want to be using the setting out surface for other tasks get a piece of plywood as a movable board that can be set aside when the table is needed for more mundane tasks such as eating.

Dimensions for the setting out table will be dependant on the type of work undertaken – medieval setting out, for example, was made on a smoothed, level floor of plaster, large enough to draw the massive windows for cathedrals.

Modern masonry companies have setting out rooms with

DEVELOPMENT

Stones do not always have square (right-angled) elevations. If this is the case there may need to be a developed section – which is the shape of the section after being (hypothetically) cut at an angle. For example a moulding that is a quadrant of a circle will with development become an elliptical curve. A developed section cut at right angles to the face will always keep one dimension the same and be extended in the other.

There is no need to get overly concerned at this point as the level of drawing skills needed in the setting out process of the exercises shown here will be kept quite basic. If this subject interests you it will be useful to get hold of one of the numerous texts on the geometry of building for simple reference to techniques.

A medieval setting-out room in Normandy demonstrated by the author for a documentary.

a table about eight foot square (the size of two readily available plywood sheets). As a good start, begin modestly with a piece of plywood about 3ft by 4ft, thick enough to remain stable and take punishment. This should have edges straight and square (in masonry terms) to allow the use of tri-squares and straight edges for geometric precision; if this is not guaranteed, securely attach a straight edge along one side to work from.

Drawing board and instruments.

Drawing Boards

A side effect of the almost universal introduction and use of computer aided design (CAD) by architects, engineers and designers is that good quality drawing office equipment is no longer needed and secondhand equipment can be had at very reasonable cost. Large upright drawing boards, or drafting machines, with parallel motion and stands are ideal for setting out and producing templets. As an easel for designing and holding information that is easily viewable while working, they become indispensable.

TECHNIQUES

The nuances of setting out for each piece will be described in the subsequent relevant chapters. Here we will cover the basic techniques and information you need to know.

The sheet for the templet should always be securely fixed to the board, not to be moved until the design work is finished, using strong (gaffer) tape or drawing (hence the name!) pins, so that there is no discrepancy in accuracy from reaffixing the sheet in the wrong position.

Line up at least one of the straight edges of the templet with the edge of the board or square them up before securing.

Dimensioning

Measure three times before final marking and cutting, to make sure the dimension is correct. Always draw lines by means of a fine point and carry them past intersections.

Whenever you are using a workshop measuring tool, be aware that the end can be out of true, as it gets abraded away on the stone. To counter any errors this can cause always use a point along the rule to start off from and, crucially, remember to subtract this from the final dimension.

Running dimensions, on the other hand, should always be taken from a single point (usually the corner), to prevent errors from accumulation.

Marking Out

Pencils should always be kept sharp. A small patch of sandpaper for fining the point can be pinned to the side of the drawing board; this will give you a chisel edge that will draw very fine lines. For a scriber the same principle of slightly flatting the point on an engineering file will let it slide into the material easier than a coned point.

Using fine-line ink pens is fine, but be aware that on impervious sheet there is a risk of smudging if it does not dry quickly. Also when marking along a straight edge you must ensure the ink does not run under the blade – straight edges and templates for use in drawing offices will have a slight bevel on the underside to prevent this.

Scribing onto Zinc

Scribing on zinc should be a smooth action, against straight edges that are held down securely with enough pressure to get a good quality line. Repeated scribing along the same line will eventually cut through the metal, which can be more accurate than using snips. Snips are used to cut into intricate shapes or mitres so the metal can be taken off in small sections.

Scribed arcs on zinc or plastic should be from a centre tapped into the material with a sharp tack for an indent into which the point can lodge securely. If the centre is off the templet attach a small piece of zinc where it should be and centre-pop this. Use two hands to describe the arc, holding the point of the dividers securely in the centre and the other end close to the scribing point. It is best to have sturdy dividers that can be locked into position for this manoeuvre.

Carefully work any excess off the edge of the zinc with an engineer's file; hold the templet flat on the table with the working area just over to prevent the sheet buckling. Take off any burrs with fine abrasive paper, then rub down all edges with wire wool.

Clean the sheet with wire wool prior to writing on it.

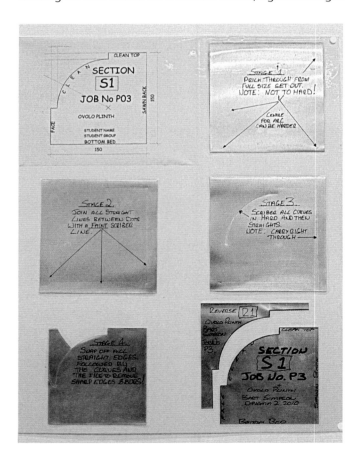

Templet instructions for students.

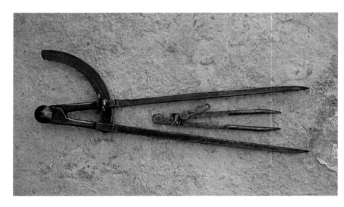

Large dividers.

TEMPLET INFORMATION

Consider how to describe a piece of stone as a two-dimensional drawing on the templet.

If you have a bedmould that effectively shows the plan of the stone then this is applied from the top, and any hidden detail or cuts underneath the stone must be marked in with a dashed line. The same goes for any section templets – the templet must show all detail even if it cannot be seen from the side it is applied to.

Mark all the relevant information – clean edges, joints and

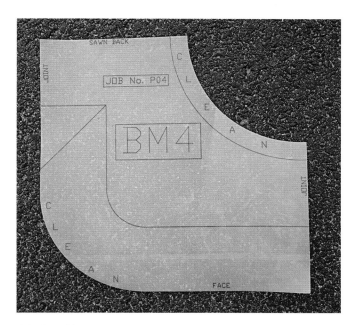

A bedmould.

so on – onto the templet with permanent marker. Match all reverses to their curves, noting on them which side they are used from and to which curve; scribe this onto zinc for permanency.

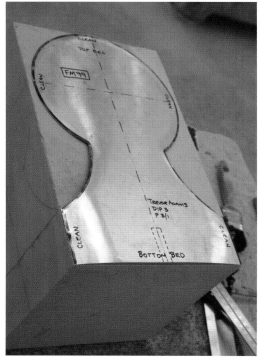

Ball finial templet applied to stone.

APPLYING A TEMPLET

Here is where the incentive for accurately sized and squared stones comes to the fore. The templet at its extremes should touch the edges of the stone when it is placed correctly on the face of the stone. Scribing or drawing around the templet will give the edges to cut to from that elevation into the stone – with minimum stone to cut off.

The intersections of planes from different faces must be marked on the stone. Items such as mitres, stops and returns are the limits of cutting in one direction. It is best to scribe all lines, as pencil will rub off quickly – but do not scribe across faces that will be exposed when the stone is fixed. Thinking of the future, setting out lines left on the sides and backs of stones will be useful to your descendants when they need to replace stone that has external edges eroded away.

Line up the edges by holding a short straightedge projecting past the stone and slide the templet to it in all planes. Then hold the templet securely and scribe along it. It is good practice to draw around on all faces using a sharp pencil first and connecting with a square all the relevant points to check the stone is square, then reapply and scribe.

Plastic film can buckle as you scribe along it, so hold down the templet near the edge being marked and use the templet as a guide – run the scriber along it, not against it.

Once all the lines are scribed in, use a sharp pencil to draw in the scribed lines so that you can see them. They will rapidly fill in with dust and become invisible so the pencil will show up better when the dust is blown or brushed off.

Get rid of all the dust on the stone. Place the templet on again, check that all is well and once you are satisfied, get to work cutting the stone.

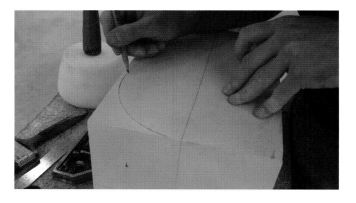

Pencilling-in a scribed line.

CHAPTER 4

USING A MALLET AND CHISEL

You will have become familiar with the feel and weight of the basic tools as you organize and prepare them. It is now time to take the first steps in stonemasonry. While it is widely held that all the theory in the world will come to naught without the practice, please do not be daunted. The greatest cathedrals in the world were designed by mere mortals.

Obviously these were men (sorry, women, but this was way back when) of tremendous vision and clarity, not to mention the bravery of building in such a new style. But they all had to start somewhere, and incredibly their first skill was to be proficient in the exercises set out in this book, using basically the same tools you are about to take up – after the exercises have been completed your foundations will be the same as theirs. Ready to build your own cathedral?

GETTING STARTED

The workshop is prepared, the stone to be worked is resting untouched on the clean dust-free, carpeted surface of the banker and you draw near the scene with tools at the ready.

Standing with legs slightly apart in front of the banker, get comfortable with shoulders relaxed. Your chisel side is angled to be closest to the stone, holding the chisel just off the centre of the chest with the mallet held about a foot away, keeping both level with the bottom of the ribcage; allowing you to strike across the body, not away from it, to bring about a consistent and accurate style.

OPPOSITE: Cleaning up the inner edge of a fillet using boaster with a brass dummy. The knowledge and expertise of how to use tools, that have changed little throughout history, can produce works of beauty and incredible accuracy.

Natural Geometry

There will be much mention of angles, levels and verticals throughout the work. This is because the stonemason must be spatially aware and capable of gauging such things as horizontal lines and equal distances accurately by sight and telling whether a line is straight with a glance. Obviously there

The author's son Rudi showing a good stance for working a stone.

will be a need to use measuring equipment and reverses, but the proficient stonemason will come to depend upon their perception of correct line, angle and level almost by instinct, resorting to the use of equipment mainly to confirm their judgment.

This ability does not rely on attaining magical levels of training but is rather a result of using the natural skills such as balance and depth perception inherent in us all. As a consequence whenever an edge is not straight, vertical, horizontal or parallel when it should be, you will sense a niggle that something is not right, which will cause you to look at it a second time to ascertain why this is so. You will develop the ability to raise issues when you are asked to comment

This hand rail in a modern restoration project of a historically important building in Dorset is an example of slipshod masonry and unskilled building control. At first glance, and to the client, this stonework looks acceptable, but it is actually very badly made. The stone is worked at the section of the joint, then it balloons out, or slopes out of true, causing the sausage-like appearance. The huge joint, apparently to allow better levelling, does not help.

on modern construction or amateur attempts at craft, as you automatically note the errors and crooked lines of supposedly good work.

Grip

The chisel should not be gripped tightly. It should be held in the loosely clenched fingers of a flattened fist (your left if right-handed), with the cutting edge protruding well past the little finger, thumb resting on the last knuckle of the first finger or, if you find it more suited for controlling the position of the chisel, against the shaft below the head of the chisel. Something that is not immediately obvious to the layperson is that at no point does the chisel hand ever rest on the

stone whilst working, as this will cause the point of strike to be restricted to where the hand can slide around comfortably on the stone. Keep off the stone.

Chisel grip.

Chisel grip.

Chisel grip.

(Traditionally French masons work with the chisel going over the top of the little finger. On occasion this manner is affected by others who wish to appear cosmopolitan, but for the moment let's stick with the way that works, unless something else really is found to be more effective for chisel grip.)

Angle of Attack

MASONED, WORKED OR CARVED?

Technically the result of these labours should be termed masoned or worked stone, but as it is easier to convey the meaning with the term 'carved' we will use it throughout the book – being aware that for some applications the term carving is for embellishment of the worked piece and not this basic stonemasonry.

There are three approximate angles that are important to the way a stonemason works; these should be determined and maintained during the working of a level draft: (1) The cutting edge of the chisel should be horizontal and (2) turned out about 30 degrees to a line drawn through your shoulders; and finally (3) the chisel should be placed with the shaft at an angle to the stone that will allow the edge to bite and efficiently remove stone, without becoming lodged in the bulk of the material.

If it is too shallow an angle the chisel will slide uselessly over, whilst one too extreme will result in the chisel locking in and not removing any stone; calculating this will become second nature quite quickly. Once you feel that the co-ordination of these three has been achieved you can take the first cut. (never use the term chip).

higher skill levels not taught here yet and should not be tried, as the second hit tends to be wasted in energy as well as cutting from a position not of your choosing.

The chisel will slide into the stone taking a direction that is the product of how hard the stone is, what angle it was hit at and the sharpness of the strike. Shards and dust will fly off (eye protection will immediately fulfil its function) and a small area of stone, clean and artificial will be created. This reshaping of a material formed millions of years ago ushers you into a journey of immense satisfaction and achievement.

Placing the chisel halfway into the line in the stone formed by the first cut, repeat this along the draft to be worked, leaning into the work without becoming overbalanced (surfer stance), limiting over-extension by moving around the stone rather than staying in one place, and always keeping an eye to the area to be cut.

THE STAGES OF CHISELLING:

1. Place the chisel at the corner and strike it so it cuts diagonally into the stone;
2. Place the chisel halfway along the back edge of the cut and strike again in the same direction;
3. Repeat to near the end of the stone, then cut in from the other corner.

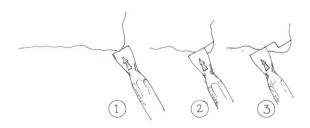

Chisel strikes

CUTTING

Start by really relaxing those shoulders and holding the tools firmly without clenching. Draw a bead on the chisel by placing the mallet head on the chisel head, comfortably bring it back to the ready position (your arm now knows which route to take) and strike sharply once, immediately repositioning for the next strike. Do not take a follow-on tap until the chisel has been repositioned. To use a double tap effectively requires

FIRST DRAFT

The result of working in this way will be a worked surface (draft) that leaves a regular serrated cut in the stone rather than a parallel edged ribbon strip. This method controls the evenness of the draft as the level of the cut is maintained by working from an easily maintained position; attempting to cut away from the body will result in over-reach and unevenness as the arms become less comfortable and controllable.

Obviously this is fine for cutting a horizontal line, but most stonemasonry will involve a bit more than this wonderful skill, so now rotate the chisel and to some extent the hand upwards around the wrist to retain the original stance but with the cut going in at 90 degrees (vertical) to the first cut.

Working across the stone perpendicular to the edge of the draft you will notice how the stone cuts easier when going into an area that has been relieved. Once the stone is removed you will end up with a serrated line running in the corner; seize the chance to practice a little accuracy by attempting to keep this line a constant shape as you work along the cut.

Once the first cuts are done the serrated line is now waste and needs to be removed. Here you may take up a wider chisel or boaster as the extra length of the cut will maintain a straighter line, while the power of the strike will be dissipated along more blade giving a softer cut from the original beat.

Rotate the end of the chisel out and rest it on the points of the serrations using the (vertical) draft level as starting point. With the chisel pointed at right angle into the corner, take a bead and strike down into the corner to shear off the waste. Then do the same from the horizontal draft, and repeat this until the draft cut is two smooth(ish!) surfaces meeting at a clean edge in the corner.

CRITICAL REFLECTION

Study your incredible intrusion into the ancient stone and demand of yourself – is this as good as I can make it? Be honest when answering. A critical attitude to the quality of one's work is an attribute every stonemason should hold onto, so run your eye down the flats and see where concentration waivered, the chisel wasn't placed correctly before striking or enthusiasm overcame control as the mallet was swung too hard, creating a chisel cut below the line, and take note.

Don't waste time attempting to jolly this draft into a reasonable condition by removing highs or adding neater cuts. In the future you should be expecting to get the first cut almost perfect in one operation. Take a break. Resolve to slow down, and breathing calmly do the whole thing again (and again) until you reach a point where you can get the chisel to cut exactly where and by how much you want it to, creating an efficient trinity of accurately placed tools, married to a controlled system of working, sufficient to the resistance of the stone.

ACHES AND PAINS

Muscles unaccustomed to this task will complain at first. This will soon pass as you achieve dexterity and the technique comes easier. The power of the strike is accomplished more by momentum than muscle; let the weight of the mallet or hammer do the work. Just get it going sharply in the right direction – physics does the rest.

The friction of the mallet moving around in your hand will probably cause blisters at first, so you may need to wear a glove while they heal (don't wear a glove on the chisel hand as it will impede control). Both your hands will eventually become callused, with blisters appearing first on the mallet hand; use an appropriate lotion to keep skin supple after work.

Before starting to work apply a barrier cream and the drying effects of stone dust will not be so marked (this is also essential when fixing stone as the lime in mortars, being quite caustic and hygroscopic in nature, can cause dry, cracked and very painful skin).

It is important to be comfortable and relaxed in order to work effectively and accurately for long periods. The strain from tensed muscle and awkward joint rotation can be considerable, resulting in short term aches and pains followed by long term injury if working positions and movement are not suitably ergonomic.

The author's daughter Lily lines up the strike, using the weight of the mallet and the length of her arm as a lever to do the work.

Take Your Time

Do not attempt to work fast – it is essential to become accurate with the chisel prior to any attempt at working quickly. Fast chiselling in untrained hands will result in errors and rough work if it is not accurate; once accuracy in cutting is achieved faster work rates will certainly follow.

Working in the Real World

Though carved stone is still mainly produced through hand-working by a banker mason, many now use disc cutters and air tools to speed up the work by getting rid of the waste quickly. It is possible to become so adept with power tools that the use of chisel is limited to the final stages of finishing or dressing in, if at all.

It may be that you start to explore this avenue, believing it to be the way ahead. We all know technology is a wonderful thing and should never be dismissed – consider the astounding advance when chisels and mallets took over from shaping stone by pounding with another stone! But just as Picasso could only produce his glorious work from knowing the established way of drawing and painting, the stonemason should only consider the use of power tools after having completing all the exercises and being fully aware of the processes you need if the power gives out.

Lathes and planing machines are sometimes used (on the same principles as joinery, though of a heavier duty) for production of columns, balusters, cornice and other straight runs of moulding, though there is still a need for masons finishing off the bits where the machine can't get into.

Dawson using air-tool to make a pineapple top.

CHAPTER 5

LETTERING

Stonemasons traditionally have a method of marking the stone, known as the mason's or banker mark, to show that the mason had produced the work – originally for the master-mason or tallyman to inspect and in calculation of wages. These marks can often be found on old stonework, usually hidden away perhaps in the joint of the stone, though some buildings that are using up stone quickly can have them in quite visible places to show how much work from individual masons has been fixed.

This was especially pertinent when many stonemasons worked together on a large project, such as a medieval cathedral or castle. The building of Beaumaris Castle in North Wales, started in 1295, employed 400 masons, and a thousand unskilled workers, as well as all the other trades; this number equaled, at that time, about 13 per cent of the adult workforce in London. In those days of general illiteracy it would be unusual to have everybody writing their name, or to even have the need to, whereas journeymen would need to apply a moniker to employment papers and invoices. A distinctive design would be the serf's equivalent of the aristocratic seal.

Typically the simplest types of mason's mark would be an arrangement of straight lines, easily and speedily cut with a flat chisel. Whilst this allows for a lot of permutations, it may be good practice to up the ante here and bring a bit of style to the work, as well as get a practical introduction to lettercutting.

OPPOSITE: Lettercutting should be an amalgamation of advanced tool skills, high quality design and art; to reach the standards necessary for work of this level takes us beyond the craft of the artisan stonemason.

DESIGNING YOUR MASON'S MARK

As an enjoyable exercise, you can design a personal mark you feel would be practical to cut into each of your stones, remembering that this will allow you to be identified through-

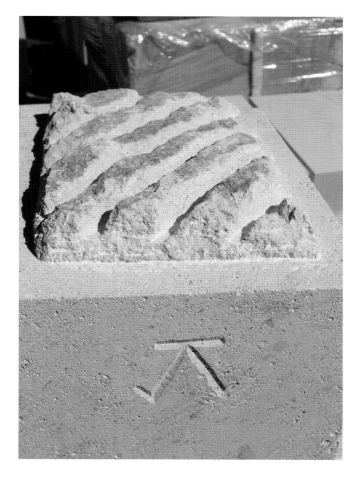

Mason's mark, on a block with a nice furrowed rock face with margin finish to the stone.

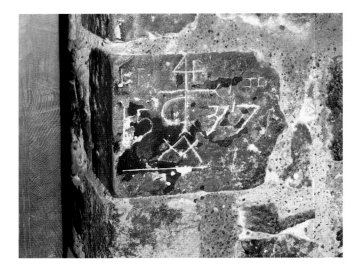

Old mason's mark (1577) in the Tower of London. These can be found in many historic buildings; the styles may differ, but all are signs of a specific mason's lodge.

ROMAN LETTERS

Trajan's Column, erected in 113 AD, is one of the most impressive monuments in the world. Standing about 35 metres high, it has a continuous frieze showing the emperor's triumph in the Dacian wars. All very interesting but what is really impressive is that the column is made up of 33-tonne marble drums and the cap stone weighs over 53 tonnes – all which had to be lifted without machinery as we know it.

   The lettering panel on the plinth is the best example of Roman Square letters, and it is on these that many of our modern type-faces are based. The bottom row is smaller than the top, allowing for perspective when being read from ground level.

Letters at base of Trajan Column.

out time so be sure of the finished article – is it neat, does it have significance and can you carve it (happily this last task will become easy once you have finished your first piece). So design your mark, from as few letters/shapes as necessary to get the desired effect when cut into stone, so bold simple shapes are good. Consider that it is important to be able to draw and execute it without too much effort.

Realistically it would take another volume to explain the design and form of letters, so consider the exercise below as a practical taster, then make it your own task or homework to create a design that suits you.

After coming up with the design the first thing to do is get the surface of the stone, where the mark is going, smooth enough to use; so work it down with chisels, sandpaper or whatever works to get a flat, mark-free area large enough for the design.

Now draw the mark on, using a sharp hard pencil or, if it is easier, have the design on paper and transfer by tracing over carbon paper.

FREEDOM OF EXPRESSION OR DISCIPLINE?

Calligraphy is a visual art, often defined by the complications of the letter. The style should be disciplined yet flowing – similar to tracery work in stone. It's all in the design.. The professional lettercarver, when producing a piece, spends most of their time drawing and spacing the letters. The cutting is simply the process of transferring this to the stone, the result being regularly sized letters perfectly set out. This approach is highly commendable and should be the foundation for anyone taking up this skill.

Conversely visiting museums and churchyards will show how the use of undisciplined design (in calligraphers' terms) can often result in beautiful, naturalistic work to rival any.

CUTTING INCISED LETTERS

Lettercutting, or lettercarving, uses tool skills well within the grasp of a competent stonemason as it is all about cutting to the line, the difference being that in lettercutting you will be attempting to produce a v-shaped furrow.

There are three types of chisel action used here: stabbing, chopping and chasing, which, as in all stoneworking, rely on precision and consistency, just on a smaller scale.

For this demonstration, we will produce a letter 'R', which can incorporate all the styles of cutting for incised letters.

Remember as this is merely a taster it will be your task to either take up further training or develop the skills independently; do not consider this formal lettercutting training in any way.

CHASING THE CURVE

Good practice is to cut curved sections first; this allows thinner parts of the letter, where the curve intersects the straight, to be in place before the fatter part is cut.

For this it will be best to use a dummy that fits into the palm of your hand to allow extra control, and a super sharp chisel that will slice through stone with little recourse to swinging the hammer. A dedicated lettercutting chisel would be a nice acquisition to use here; practically though a slender bladed half-inch masonry will accomplish the task.

Starting at the thinnest part the task is to cut the finest line possible down the centre of the form, using gentle quick, repetitive taps on the chisel, and twisting the hand around to keep a consistent angle of about 60 degrees from the horizontal for the long axis of the blade; the chisel is brought down close to horizontal to reduce the depth of the cut, and a shallow furrow is formed.

Twist the blade to the opposite angle and repeat the action but cutting progressively deeper at the wider part of the form, coming up shallower as it narrows again. The chisel hand must never rest on the stone as this will restrict movement and curve.

As cutting out of stone can result in spalling, it is necessary as the curve progresses to rotate the head of the chisel to keep the blade slicing down through the surface. Another aspect is that the cut may come back towards you, and so it is best to get some practice in rotating the strike direction without physically moving around the stone.

Finish the joining cuts off inside the body of the upright or straight section as desired; the tapered point requires that the chisel is brought up almost to the vertical to get the gentle slope.

STRAIGHT LINES

Stabbing

For straight lines, especially in small formal lettering such as military gravestones, stabbing was used to get a simple straight-sided form.

Take a chisel with a blade almost the length of the form (on larger letters you may need to do this more than once along the letter), and place it on the centre line with the shaft vertical. Give it a sharp strike downwards; this puts a line of possible fracture running to the centre and bottom of the letter.

Now place the chisel on the limit of the letter and tilt it slightly out of vertical, so the direction is just inside the letter. Give this a sharp strike and the cut will deflect perfectly into the centre. Do this on the other side and the result will be a regular V channel that just needs tidying up at the ends or embellishing with serifs.

Chopping

The letter here is shown without perfectly straight lines. Not many letters that are hand drawn tend to have them, and so while the stabbing can be used to get the bulk out of the centre there will need to be some shaping of the sides and the forming of serifs.

SNEAK PREVIEW

Sometimes it is hard, in paler stones, to determine how the letter will look when all is cleaned up. Lay a piece of paper on top of the letter(ing) and with a soft pencil or wax crayon shade across with even pressure. This will produce a rubbing that gives a perfect picture of the letter, highlighting any discrepancies in the edges. It is also a useful way to keep records of any lettering.

This can be accomplished by chasing or, as many people prefer, by chopping out and then some gentle work to the finish; these decisions will be made based on the size of the letter and the cutting characteristics of the stone.

Chopping is the same as cutting a draft on stone, except here an angled draft is required, and as the letter widens the cut gets deeper.

The centre is relieved by stabbing or chopping a channel on the centre and then to the line, with the mess in the bottom cut by rotating the chisel and stabbing down to make a smooth even line. Alternatively get the chisel facing along the channel and chase along the length of the cut to clean up the bottom. Make sure to keep the angle the same as the wall of the V and do not dig in as you move along. Where a line

Raised lettering.

intersects into another the aim is to cut cleanly through the channel to reveal a neat V section in the side of the letter wall.

Serifs are cut by placing the chisel at the end of the letter and rotating the head round while striking ever more gently till the blade rolls onto the point to get the sharp curve; practise this to get regular shapes and quality.

The top of the letter is finished off by starting at the point of the serif (angled to 60 degrees) and chasing down towards the other serif point; as the corner of the chisel gets to the bottom of the furrow the head is rotated over to get the blade cutting in for the top side of the opposite serif.

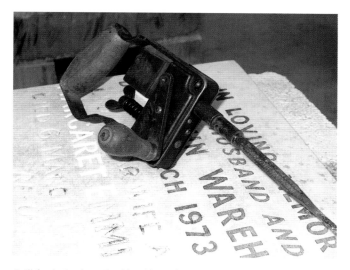

Drill for holes for raised lead lettering.

Raised lettering (author).

RAISED LETTERING

Not all lettering is incised – it can be raised, with the background cut back.

Raised lead letters were another technique, where the lead was beaten flat onto the stone, keying itself into holes drilled in the face at appropriate spaces and then the letter was cut out with a knife.

Dummy chisels for lettering and carving.

CHAPTER 6

BONING IN AND WORKING A FLAT SURFACE

The natural formation of sedimentary stone can produce surprisingly regular blocks, the bed height dictating the height of the courses in the building and often having a planar surface at right angles to the bed, giving the opportunity to use it as the finished surface to a wall. Whilst this can produce masonry of good quality it is, no matter how even the presented surface, recognized as rubble stone work and is the craft of the walling mason who is often, erroneously, called a stonemason.

To produce worked masonry and thus dictate the quality and style of the finished surface requires the training and experience of the stonemason, even if only to produce a flat surface. Taking a block of quarried stone and levelling the surface to a true even plane is the first task for all stonemasons.

Facing an unprocessed block of stone (to the layman: boulder) with no regular surface the mason must be able to define a precise starting point by boning in the (rough faced) stone. The common but difficult task is to create a flat surface either to use as part of a wall or to begin the creation of a precisely shaped element of a structure. This is seemingly the simplest and most basic exercise for a stonemason, yet in terms of accomplishment the most demanding as here the techniques and skills that define the artisan will be born.

RIGHT: Random-coursed rubble stonework in red sandstone offers a fine example of the builder mason's work. The dressed details provide the necessary tying off of edges and surfaces for shedding water.

OPPOSITE: Obelisk at Luxor in Egypt; a single piece of stone that was worked in the same manner as all the stonemasonry in this book. The incredible feat of bringing such a monolith to a set of flat regular surfaces is simply the techniques described within on a grand scale – a future project. (Photo: Dawn Fischer)

HAND WROUGHT BEAUTY

Craft stonemasonry is concerned with traditional design, where the surface of a masonry structure reads as well as the decorative elements. The only time a stone should be used un-worked from the sawyer is when that (cut) face is within the confines of a structure, and only if a dead flat surface is appropriate.

The façade of a building becomes alive with the tale told by the strokes of a chisel or the hand dragged finish. There is no life in the insipid blandness of a machined surface, where even seemingly plain stones are worked and termed as a rubbed finish.

Using well-wrought stonework in a building gives it texture and depth, allowing time and the elements to prank it with the patination of deserved age. As with Botox injections and cosmetic plastic surgery, unnaturally smooth areas in a weathered visage usually jar and often repulse.

For this exercise we are romantically assuming that perhaps just one stone is needed and it is hiding within the confines of an irregular block rescued from a tumbled wall or lurking in the corner of the workshop, economics and time conspiring to make it necessary and practical to carry out the work by hand.

This stone (continuing our thread) must be fitted into an existing wall of hand-worked ashlar, so logistical requirements are that the face must be perfect, the joints worked in the manner of the original and the depth (distance from front to back) equal to support the stone and fill the available space.

For each exercise in the book, unless there are specific tools or consumables needed for the task, which will be mentioned at the beginning, just open your toolbox, unroll the tool-roll and take up the tool as stated in the text. For this task boning blocks, four identically sized cubes of hard wood

about 2in/50mm a side, and two straight edges are the supplementary tools.

BONING IN THE BLOCK

With dimensions and volume considered, set the stone on the banker with the most appropriate face to work laid up roughly parallel with the floor and supported by wedges to prevent movement whilst working. Remember that the incessant striking will cause enough movement for the stone to rock off wedges, so if possible fix any supports securely.

Take up the mallet and chisel and start by cutting a check into one corner, working carefully whilst taking out the stone. The result should be an adequately sized niche (2in/5cm

Cutting-in checks for the blocks to sit in.

Set-up for boning in. The chisels hold the straight edges in position, allowing the eying-up to be hands free.

square) with three worked faces for the boning block to rest in. Attempt to make this area's base in the same plane as the proposed flat surface, but don't get too caught up in worrying about the accuracy yet – it will come.

Then carry out the same operation at the other corner of the long side until two niches are cut at either end of one side of the stone. Now confirm they are level with each other by checking the straight edge sits tightly on each block. It is important the platform (levels) are connected by an imaginary straight line that lies below the rough surface of the stone.

Note: It is best to start a little high or to ensure that there is enough body of stone to allow a reworking of these, to gain some pertinent practice or if things go awry.

Gauging by eye, form two more platforms in the opposite corners and at approximately the same height, to set the stage for boning in.

Place the blocks on the platforms and set a straight edge across the longest sides, held in place by using a chisel as a strut; then stand back, and by eyeing across the top of the front straight edge to the rear straight edge you will be able to judge how far out the imaginary straight line at the back is.

It is now the time to bring these edges into line by shaving away the appropriate platform, until the edges line up perfectly by sight along their whole length and the blocks sit evenly beneath the straight edges. Note that whenever the term perfect is used, do not assume this is the just for one

particular stage; it is the level to aspire to in all aspects of your work. The crucial point here is that even a slight discrepancy at the beginning will be exaggerated throughout the finished stone and possibly cause it to be unusable.

Once alignment of the straight edges is achieved you will now cut four flats into the stone on the same plane or level. These will be at the finished level of your stone's flat surface, so before the next stage cast a critical eye across the rough stone between them to make sure that all the rough stone in the centre is above the (imagined) level. If it is not, and you work the stone down in this state, all the stages will need to be carried out again, whereas boning in the corners to a slightly lower level is relatively easy.

Cutting the Drafts

Guided by an imaginary line connecting the two corner flats, cut a rough draft to join them using a 1inch chisel, taking care to keep above the level, chopping horizontally from one end to the other then rotating the chisel 90 degrees and repeating to produce a step along the edge, as practised already. Continue in this manner until the draft is perceived to be nearing the required level. Here is where the measure of the stones' capacity to take a sharp face is gained as you find out how the stone reacts to a chisel, constantly adjusting the cut to suit.

Remember to cut in a short section from the other end to prevent the corner spalling.

Checking Surface Levels

Placing the straight edge across this draft determine the areas of stone to be removed to complete a flat draft. Accomplish this by working off the highest spots shown by the straight edge, bringing the level to a whole, all the while becoming more attuned to the working qualities of the stone as you go.

As the draft progresses to the level required, cutting should become more accurate and regular until finally the draft is finely tooled and flush with the straight edge along its length (and width).

All the way through this (and every other flat surface worked) it is essential to be able to identify the high spots efficiently. They will gradually become harder to discern as the variations in the surface blend into the background when you are getting closer to the final cut, so get a bit of help by

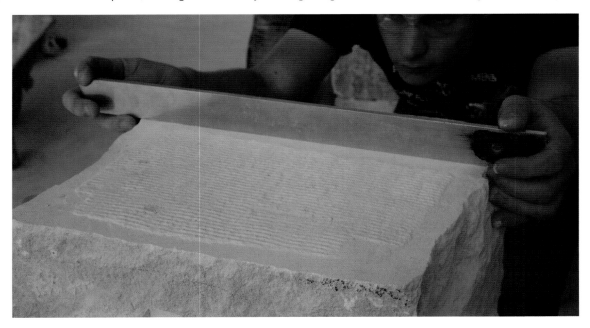

Checking for level.

rubbing pencil lead along the straight edge, then sliding this to and fro on the stone – the high spots can be discerned by dark lines left on the stone. Experience will show the amount of cutting required to remove high spots effectively; a rule of thumb is to start at a point midway between the high and low areas, then work this across the area to beyond the high spot. As long as the draft is perfectly straight, at this stage a slight overworking of the level will not be detrimental to the finished stone; the opposite levels can be easily boned in again.

Now cut the drafts to the shorter sides in the same manner to join up the first drafts. If the stone is a regular block these may give a rectangular plan, but do not be overly concerned if they are not square – a flat surface is the goal here so just keep an eye on the levels, checking all the time that opposite sides are level to each other.

CONSIDER: effectiveness in accurately chiselling stone comes through practice, so it may be prudent to work all the drafts around the stone to a close level first before tackling the final finishing surface. This gives, depending on the size and stone, a few more hours getting tool handling correct before absolute accuracy is necessary.

PUNCHING AND CLAWING

Once all the drafts are worked to the same level the rough uneven stone left in the centre is ready to be removed by opening hostilities with another set of tools: the punching or

Punching off surface. (Photo: Richard Mortimer)

Hammer and punch.

Completed drafts. (Photo: Richard Mortimer)

pitching hammer and punch (traditionally for softer stones) or a point that takes care of harder stones.

Taking these heavy, brutal tools in hand you may feel that their purpose is one of crudity and bluster. Dismiss this thought immediately by considering that some of the world's greatest sculpture was produced wholly by this combination, and now it is time to discover another facet of the stonemason's craft – effective punching.

Working next to and parallel to the shortest side draft start to cut a furrow with the punch across the stone from one long draft to the other. Place the point of the chisel into the stone and strike; the result will be a radiating break ending up with a small cut in the face of the rough; by adjusting the angle and direction of strike a roughly straight line is kept to and monitored using straight edge to a regular height approximately 5mm/2in higher than the drafts.

It is best not to continue right across the stone as you could spall the edge with the power of the strike if there is not enough body to carry the energy. So always do the last section by working into the body of the stone (towards the middle); this is the rule for all stone working unless trying to knock off edges – as we will cover in pitching.

Notice how different this feels from the almost ethereal striking of a mallet. Forearms ache, the grip stiffens and the hazards of striking the (tiny) end of a punch nestling in your hand with the swung weight of a three pound hammer through its inch square face become apparent! The skin of every stonemason has been as easily wiped off flesh as tissue-paper when hit by this hammer, so as the blood wells to the surface and your knuckle throbs. Sprinkle some handy stone dust on it to soak up the flow and smile (painfully) at the ancient axiom: 'You're not a stonemason 'til you've removed enough skin from your hand to make a mason's apron'.

Repeat this (the cutting, not the injury!) across the length of the stone at approximately 25mm intervals (this distance will be determined by how the stone comes away, if a ridge is left that is significantly higher than the furrows, close up the distance) to result in an accurate but rough, regularly coursed panel between the drafts that resembles a ploughed field.

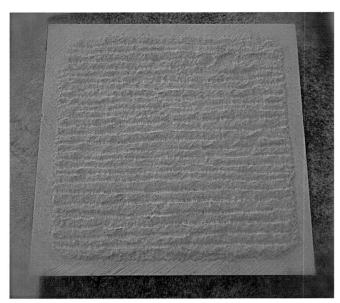

Punched surface. (Photo: Richard Mortimer)

Claw

Now taking up the mallet again and using the 2in patent claw, work regular drafts over this landscape removing the large punch lines to finish barely higher than the edge drafts to bring the whole undrafted face, as always monitoring with straight edge, to a level regular combed finish. Note how the claw is effectively a set of small punches that do not dig in as much as a chisel of comparable width allowing effective but gentle removal of stone.

Having used chisel, punch and claw you will be starting to appreciate the different ways that these attack and remove stone, the punch concentrating power to a small area that

Claw-tools. Top is TCT, while the rest are patent claw with replaceable blades; the bottom blade has been ground to give a bullnose cut.

Clawing off the punched surface to a level close to the drafts.

Clawing.

spreads the force out in a radial style, the claw which plucks the stone spreading the power to remove material quickly and the chisel with its shearing ability to produce smooth regular surface finish from the hard unforgiving structure of the rock.

Here is where the realization that physics and, more specifically, mechanics are dictating how efficient and precise the work you undertake is: the discipline of swinging a mallet head to add energy to the strike will show that to be able to lift and use such a weight all day, little actual muscle strength is required here, rather let the momentum take charge to drop the weight using the lever that is your forearm and the tool shaft; this way the strike remains constant and controlled through skill rather than the obvious progression that if you used pure muscle to hit the chisel day in day out your muscles (and the strike) will become stronger resulting in extra

The clawed surface.

care needed as brute force fights with fluidity and accuracy so relax, get in charge and let the tools do the work.

BOASTING THE FLAT SURFACE

The stone now needs to be given an even face by removing all the conflicting marks of chisel, punch and claw. This is accomplished by using a boaster, which allows the power of the strike to be spread along a wide blade, giving – when used on a prepared (clawed) surface – a precise well-controlled cut. In the hands of your future (experienced) self and on the right stone this chisel will allow you to produce a microscopically flat surface that to all intents and purposes is polished by the chisel, so prepare for this day by picking up mallet and boaster and setting to.

Boasting away from the corner at an angle of 45° to set a line from the two drafts.

At the juncture of two drafts place the chisel across the corner on the smooth drafts, and with care start working into the roughened clawed area. The premise here is that the draft is used to set the bolster at the right height, and it is a simple task to carry this level into the uncharted regions of the central area.

Now comes discipline, as the flatness of surface is to be the result of always cutting at the same height with the same power and frequency as you move across the stone. There are

Boasting parallel with the draft.

Checking across diagonals to prevent the surface twisting out of true. This is quite an important level to check as the flat surface can occasionally be straight along the outer drafts, but a small discrepancy can manifest in an out of true plane.

two methods that are commonly used: the first is to work in straight lines parallel to a draft; the other option is to work diagonally across the stone from draft to draft.

Boasting working diagonally.

The parallel straight lines method can be effective in the right hands, but the chances of error are greater as the edge of the chisel not resting on the draft can move into the stone face, with any error becoming magnified the further away from the draft. So whilst it may be taught as a traditional method, the finish can be ugly and irregular; it does require a level of skill that needs to be in place for a stonemason, but it will only be proposed in this book for long strips that do not allow the diagonal method to be used.

The Diagonal Method

The effective way is to start in the corner and, working across the drafts, start to progress over the surface with a good regular line of tool cuts giving the impression of 'hatching' the face. This method also allows you to use straight edges to monitor and check the accuracy of the cuts to the two drafts that start at the corner, rather than only being able to check effectively when the work is finished.

Boasted surface. (Photo: Richard Mortimer)

Once the surface has been worked over, it is termed boasted, either ready to use or perhaps another finish should now be considered if it is to be used as part of a wall that has a particular working to the face. There are two terms that may be used for a finish in the same vein as boasting: droved work is made with a wider chisel (2.5in/65mm) without attempting to keep the (cut) lines continuous; tooled work is made using a chisel 4in/100mm wide and running the cuts in continuous lines across the stone.

THE FINISHED SURFACE TO USE

To finish the stone off you must decide on the style and quality of the surface to end with. It seems obvious that this surface should match or at least blend in gracefully with the wall where it will be fitted, but it is depressingly common to see a sawn stone in a hand tooled wall, without even the grace to have the saw-lash rubbed out, or even with a crude attempt at 'weathering' hacked with whatever heavy object falls to hand.

Cut by a stonemason with the same tools as yours, the surface of the original will often have all the evidence of process staring you in the face, so use this to work out how to get a suitable finish on your stone. Measure the width of the chisel cuts, look at the frequency (cuts per inch), depth and importantly the direction of cut. Then, if you feel confident enough to produce a matching surface carry on; if not get a spare piece of stone to practise on until you get it right.

HAND WORKED v MACHINE CUT JOINTS

It is common practice now to replace masonry with blocks straight from the saw; possibly the face may be worked over by hand. The joints will be dead flat and true – a tactful way of calling it characterless.

Obviously since the first teeth were cut into a metal blade and saws became strong enough, a saw could be pulled through stone. Even more so in this day and age of cheap ubiquitous manufacturing, saw technology allows a regular flat surface to be cut without the need for working by hand.

As stone weathers back from the original worked surface, joints in the hand-worked stone can become more irregular – reasonably the mason would not have wasted time getting a perfect finish on surfaces that would be set in the wall and unseen. So effectively only the edge up to the arris will have needed to be exact, to give a precise edge for the jointing, behind which the stone will still have been worked to the exact dimensions; but as long as there were no pressure (pinch) spots, there was no need to produce a perfect surface. This incidentally can aid in the stability of the construction as the contours of the surface will provide a key for the mortar.

Unfortunately with sawn stone the jointed surface, flat and even from front to back, will over time present a jarring dissimilarity with the movement and irregularity of its original neighbour.

CHAPTER 7

SQUARING A BLOCK

Once the first side of the block has been worked to a dead flat surface it is now ready to turn into a usable stone suitable for insertion into a wall, or as a blank to produce a moulded piece from. The requirement here is to make a six-sided rectangular block of stone by cutting the rough sides back square from the face and then working another flat surface parallel to the first.

So first mark out on the surface the dimensions of this side, gauging by eye into the shape of the stone that there is enough body to allow the cutting of the four sides. Remember that if this was going into a wall as an ashlar block it would not need to be perfect at the back, the flat surfaces needing only to be as deep as the thickness of that part of the wall, whereas this particular block will, once that stage is reached, have a perfect back put on – all good practices.

MARKING UP THE STONE

As this is to be a moulded piece, now would be the usual time to apply the moulds and templets to the top and sides, to figure out the orientation of the stone (keeping in mind the correct way to lay stone with regard to the natural beds in the stone) so that it is possible to mark up the block that this needs to be cut from.

Scribe the lines of the rectangle into the face and run a sharpened pencil in the scribe line to help it show up clearly. Do not be tempted to make the shape slightly larger than you need just in case you go wrong, as error is not an option.

OPPOSITE: Beautifully worked tooled finish to the surface of a stone, aging gracefully with a pranking of lichen adding to its patina. The application of a few deft strokes with the chisel can lift a plain surface to the sublime.

Everything from now on relies on precision with a chisel – something you should be capable of by now (always have a practice on a dead bit of the stone if you need to reacquaint yourself with the stone's attributes – a couple of minutes should suffice).

PITCHING

Before the chiselling starts though, you must become au fait with another technique: pitching off.

Observe the pitcher. It is a stolid tool without the graceful

THE QUARRYMAN

One of the problems with stone is that it is very dense (which translates as quite heavy for its volume) so moving it about, as we have remarked, is a task to be minimized. The medieval ideal, when building, would be to get as much work done at the quarry as possible. This is the same today, but we use the saws in the processing.

The quarrymen would be given orders to square up the stone into blocks before transporting it to the stonemasonry workshops. The master mason would know the bed heights of the available stone and order accordingly, specifying an area or run of stone so the quarry could fill it economically by using available lengths of the correct height; this is evident in the varying lengths of stone in ashlar wall.

The limited masonry skills of the quarrymen would not have been a concern, as the stonemasons would dress the faces of stones that would be viewed, and much interior work would be covered with plaster.

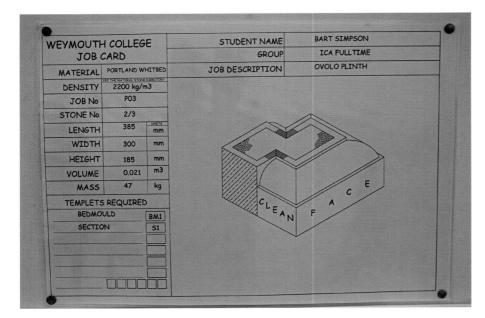

WEYMOUTH COLLEGE JOB CARD		STUDENT NAME	BART SIMPSON
		GROUP	ICA FULLTIME
MATERIAL	PORTLAND WHITBED	JOB DESCRIPTION	OVOLO PLINTH
DENSITY	2200 kg/m3 SEE THE NATURAL STONE DIRECTORY		
JOB No	P03		
STONE No	2/3		
LENGTH	385	mm UNITS	
WIDTH	300	mm	
HEIGHT	185	mm	
VOLUME	0.021	m3	
MASS	47	kg	
TEMPLETS REQUIRED			
BEDMOULD	BM1		
SECTION	S1		

Jobcard for students to use when producing a complete stone, similar to those used in industry, giving all the relevant information required to finish the stone. The sketch of the piece accurately sets out how the finished piece should look – handy when there are lots of stones sat in the yard waiting to be fixed.

lines of a chisel. In the hammer-head style, it is only efficient with the mashing or pitching hammer (mallets are not to be used on this). This one has a wide edge for this application (narrower pitchers are used for delicate pitching). It has an unusual blade that is not about cutting but the use and direction of power. The pitcher relies on the force of the hammer strike transmitting down through the point of contact with the stone, by the shallow angled edge, into the stone and setting up a line of fracture that will break off a chunk of the stone.

How the power is harnessed, and the direction of the fracture, relies on the crystalline structure of the stone and the angle at which it is struck. Harder stones with fine structure will break cleanly as the force is pushed down in a narrow

Pitching to the line. The tool is placed and held almost perpendicular to the surface, then struck clean and hard.

The pitcher. (Photo: Richard Mortimer)

area; softer stones absorb the energy and dissipate it so the pitch is not so efficient.

To get a feel for how a particular stone will react to all this aggression, choose an area with a flat surface that is not going to be kept; scribe and clean the dust from a straight line. Place the pitcher on the stone past the line and draw the tip of the blade back into the groove. Then holding it firmly just off 90

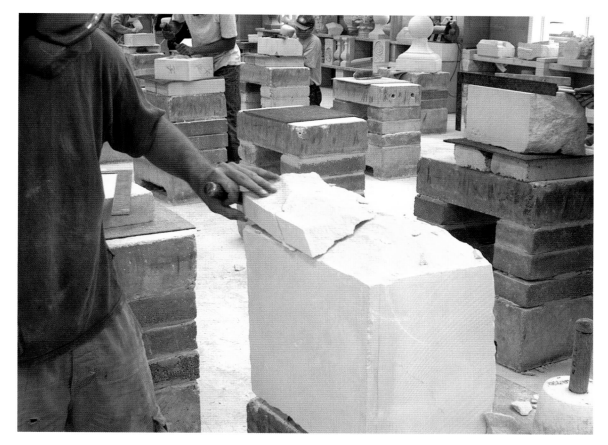

degrees strike sharply with the hammer, getting the lines of force to travel down through the chisel.

The dramatic result will be a sharp edge (arris) of the blade width starting at the scribed line, and – depending on the stone – a roughly scallop shaped chunk will fall out. In a good pitch the angle it makes with the face should be just above square with the clean face. Don't fret about getting the best result straightaway; the confidence to hit into the body of the stone is difficult at first. Experience and practice will result in ever better accuracy, which means less stone to work down to a flat surface.

So now you understand how, with controlled pitching, it is possible to remove incredible amounts of stone, set to your block and work along the scribed line on one of the long edges with the pitcher, always setting the sharp edge snugly into the groove. Wiping the blade clean after every pitch is a good habit to get into, as dust compacted on the edge will prevent it sitting in the groove properly and adversely cushion the blow as well.

THE OTHER SIDE

With the edge now holding a nice clean arris, use this as the level for cutting in a draft at right angles to the original face. Move around the stone into a comfortable position and, depending on the stone and preference, use either an inch or half-inch chisel to cut down along the arris until a draft is cut. Check progress with a sliding square set to an appropriate depth (always make sure the adjusting screw is tight, as slightest movement of the loose blade whilst checking can approve errors that will create and generate problems).

This draft is the starting point for working a flat surface into the next face, so take time and work methodically to make it good. The (pitched) arris should be perfectly straight so the only checking at this point is on every couple of chisel cuts with the sliding square.

It will be noticed that the position of the stone has not been mentioned for this section, so if possible it would be good to get this new surface up on top as this is the way we have been working so far. It will also make access to the stone easier.

Don't despair if the stone has to stay in position, however, as it is always good to practise working vertically, or more correctly in this case downhand. If you persevere, to cut a vertical draft adopt a slightly twisted stance and striking downwards, with the chisel moving down the centre of your body; you will need to hunch a bit to be able to see over your hands to where the chisel is placed.

The tactic for this draft is to rough out away from the finished level with your chisel, or if the stone allows it by pitching off chunks to get a head start.

Drape the large square down this line from the first face through the arris draft and draw a line in with pencil (obviously this is not going to be exact but will give a rough indication of the required level). Then start cutting in this new draft checking all the while with squares until it runs true.

Moving around to the other side of this flat to be, you will immediately perceive that you started on the previous draft because it was the most comfortable way to work in this plane, so this draft is going to be more awkward to do, as your body will feel like it is working unnaturally until you get into the swing of striking from the 'wrong' side. The resulting three drafts will once again surround a rocky interior. There is still the draft nearest the floor to cut; best practice is to rough this draft in without getting too finicky as it will be difficult to get good access at this stage.

Now carry out punching, clawing and boasting as before, once again remembering never to strike out of the body of stone over an edge, as you could lose too much material. Trimming in the bottom draft will require almost kneeling or lying down to get the correct position. Some squatting will be okay and probably necessary, as long as you do not overbalance, as you work horizontally cutting up into the stone, whilst wistfully thinking of adjustable height bankers or hoists that easily manoeuvre stone.

BACK TO THE HORIZONTAL

As mentioned the stone could be moved, and if you (unsurprisingly) feel after achieving this one face that this is plenty enough practice on vertical surfaces for the moment, get the

Punching the vertical.

Boasting.

stone shifted so each face being worked on is upwards. The downside of this is that pitching will be from facing the stone rather than downhand; but a practice that occasionally ends in tears will have an incentive to make perfect, as you will notice how it is more difficult to strike the pitcher hard at 90 degrees to your body.

The benefit (of turning the stone) is that boning blocks can be used for finding the levels of the drafts by cutting in checks before drafts and using the square to verify.

Carry out this task on the other sides of the stone, and by the time you get to the back your experience in the tools skills of a flat surface will be at least twice that of the average student, who only has to produce one or two during training. Once you have done two sides this also allows you to cheat and get it finished by sawing, but obviously there will be the issues of getting it to a suitable sawyer as well as time and cost, so perhaps just carry on until it is ready to apply the templet.

USING POWER TOOLS

Power tools such as disc-cutters or adapted chain-saws are used to great effect for cutting off and preparing stone ready for the finishing strokes of hand tools. Used correctly they are efficient, cost effective and can allow more stonemasonry to be carried out by a one man band; in a big project the task of turning every rough boulder into a honed block by hand,

enjoyable as it is, could become a bit of a drudge. So don't consider them tools unworthy of the artisan.

Blade technology allows easy access to all manner of cutting and grinding discs, so choose the one most suited for the stone being worked. Get into the correct protective gear and figure out how to be effective in following a scribed line when engulfed in clouds of dust, then set to. (If you have no extractors go outside, but remember this is a very unsociable practice!

Cutting by Disc

Holding the tool in a relaxed but controlling grip, stand so that you do not overbalance as you move forward and start cutting along the outside of the line, working down in slow incremental passes. Cutting too deeply too quickly can jam the blade and result in injury. To be cautious is wise, as doing anything carelessly or too hastily with these machines will result in incidents that range from cutting the wrong part of the stone all the way up to hideous injury.

Important notice: It is a requirement by law that people who use these power tools should get some training. Unless you have somebody experienced showing you how to use them, they are not recommended here and will not be mentioned again for stone cutting unless they are the best possible option, which occasionally they are.

The block is now ready to be masoned.

POWER TOOLS SAFETY INSTRUCTIONS

Every tool has its rules and precautions that apply to it. In the case of power tools, many of these are the same for each tool every time. Learn these by heart and you will always be off to a safe start.

Good Practice

Always read, understand and follow the instruction manual before attempting to use any power tool in any way. Also read the nameplate information and follow the warning labels on the tool itself.

Clothes and Equipment

- Always wear safety goggles or safety glasses with side shields. Use a dust mask for dusty operations, and wear hearing protection if you will be using the tool for an extended period of time.
- Dress appropriately and safely. No loose-fitting clothing, no neckties, no jewellery, no dangling objects of any kind. Long hair must be tied back out of way.
- Safety footwear must be worn and laces tied.

Work Area

- Make sure your work area is neat and clean and free of any debris that might get in your way or be ignited by hot tools, chips or sparks.
- Make sure your work area is well illuminated and that there is adequate ventilation to prevent the build up of dust that could obscure work.

Using the Tool

- Ensure that the machine is not damaged, cables are clean of cuts and that plugs are attached correctly. All machinery must have a current Portable Appliance Test (PAT) certificate.
- Before you plug in any power tool, make sure the power switch is off.
- Be sure all appropriate guards are in place, secure and working.
- Always turn off and unplug the tool before you make any adjustments or change accessories.
- Allow the machine to come to a stop by its own volition – do not jam it into scaffold boards etc to slow it down.
- Always use the correct spanners/keys for changing and tightening blades/bits.

Essential Precautions

About the Tool

- Always use the right tool for the right job. No substitutions allowed!
- Never use any accessory except those specifically supplied or recommended by the manufacturer. They should be described in the tool's instruction manual.
- Use the correct blade for the material you are cutting.
- Need an extension cord? Make sure it is a heavy-duty cord and always use 110V outside. Do not use 24V cords outside.
- Never use a tool that is damaged or malfunctioning in any way.
- Make sure cutters or blades are clean, sharp and securely in place. Never use bent, broken, or warped blades or cutters.

BSSH WORKING PRACTICE

- Never use power tools in wet or damp conditions.
- Never use power tools if you are tired, sick, distracted, or under the influence of drugs or alcohol.
- When using hand-held power tools, always keep a firm grip with both hands. Losing control creates a hazardous situation.
- Do not use any tool that is too heavy for you to easily control.
- Ensure that the material to be cut is secure and cutting will not damage other material. Do not use hands or feet to hold material in position.
- Never overreach when using a power tool. Stay firmly planted on both feet.
- Never rush what you are doing. Always pay close attention. Don't let anything distract you. Think ahead!
- Always unplug, clean and store the tool in a safe, dry place when you are finished using it.

CHAPTER 8

THE OVOLO

TEMPLETS AND MOULDS

In a normal workshop situation the moulds and templets for a stone would have been made by now and the stone cut to size. Here this exercise is using what stone is available, so the templet will be made to fit the stone you have ended up with.

Assuming this, the moulding will be an ovolo carried around three sides of the stone creating two external mitres, a return and break (with internal mitre) and a stop. For simplicity's sake the dimensions of the ovolo will be on a 5in/125mm centre with ½in/15mm fillets as set out in the diagram.

There will need to be a bedmould (the view of the finished stone looking down from above) and a templet (the section through the moulding) with the dimensions above.

The bedmould can be drawn to scale on paper as the dimensions of this can be transferred direct to the stone to coincide with the application of the templet.

The Templet

Before starting to make the templet it is assumed that the traditional method of using zinc will be followed; alternatively if preferred, plastic film is easier to work using scissors and knives.

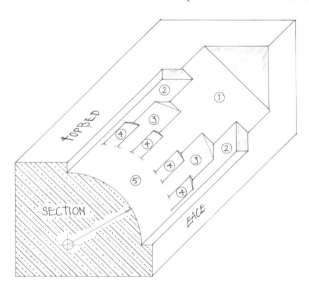

This shows the section of the ovolo and the sequence of the chamfers for cutting the curve.

OPPOSITE: The first mitre in the stonemason's journey. Up to this point the stone could have been shaped with a planer machine, but here the stonemason goes beyond modern technology and links to artisans throughout time, for they would all have been conversant with techniques such as this.

Scribing out templet onto zinc before removing excess sheet.

Set out the design of the ovolo and fillets onto a squared piece of zinc sheet using squares, scriber and dividers. Once the design is correct, drag the scriber against a metal edge through the straight lines, and similarly score the arc of the ovolo using the dividers with the intent of cutting through the sheet enough that the waste will come away with a bit of manipulating, leaving the profile of the section. It may be necessary to fettle in the edges using a fine file, wet/dry abrasive paper and wire wool to get perfect edges.

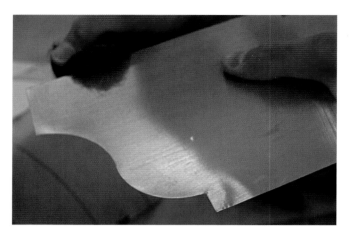

Cleaning up the edge with wire-wool.

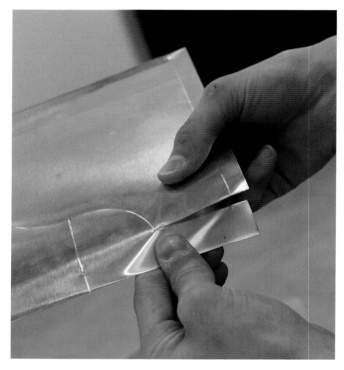

Removing waste by twisting off, so it breaks along the scribed line.

Now put a hole in so it can be hung up for safekeeping and mark it up with a permanent marker pen (as shown in the diagram). It is essential to get into the habit of putting identifying marks, dimensions and all order information on each templet, as their number will mount up and particular ones can get lost or muddled. Tin-snips are excellent for cutting zinc sheet in most applications, though sharp knives and the setting out tools can be as adequate.

Plastic film templets are marked up in the same way and cut to shape with scissors, but can be stored rolled up in tubes as mentioned.

Marking Up the Stone

Transferring the design information onto the stone is straightforward if accurate placement of the templets is carried out, so line up the templet with the corner of the stone, use an edge to make sure it sits in line with the top and front and, holding it securely, scribe around the edge. Place the templet at the other side and scribe round. The templet must be applied to every corner and scribed round as this will allow the edges of the moulding to be carried through; you may not actually cut to the end of the stone but it is important to have these guides in place.

Then set out the dimensions of the bedmould onto the stone using squares and pencil for construction lines before using the scriber to finish (not forgetting to fill in the lines with a pencil so that lines are easily visible even after brushing).

Follow the lines around to the intersections where the return breaks the side as this will need to have the section scribed on as well.

MAKING THE OVOLO

This first section is to be the moulding that runs along the side without any breaks, so the true section can be seen and worked to before attempting mitres and returns.

In cutting curved sections in stone the basic method is to

make drafts (known as chamfers in this context) that contact with the curve tangentially, then round off the edges to form the curve, so the more chamfers the easier it is to get to the finished surface. During this the fillets will also be cut in – apart from a few exceptions all mouldings tend to have fillets to their edges – they provide a frame that helps to delineate the detail, which handily gives the stonemason good levels to work from as well.

The first chamfer is set out by scribing a 45 degree line from the top of the stone to the side, just touching the outer edge of the quadrant; do this at both ends of the stone. Then joining the ends together will enclose a triangular sectioned piece of stone, the edges of which are parallel with the fillet lines.

With the pitcher placed in the scribe above the chamfer line knock the top of the triangle off at both ends. Then using ½in chisel chop in drafts along the two lines bounding the chamfer; this must be away from the finished surface and is to relieve the arris (prevent the edge spalling) by containing any strikes within the bulk of waste material. The result will be similar to the same stage of working a flat surface, except in order to remove the bulk it is best to claw down to the chopped-in drafts. Next cut in a perfect draft along the tangential line at each end. Now using the scribed lines work along the chamfer by cutting to the line on each (long) side, then working a flat across and ending up by fine boasting the surface.

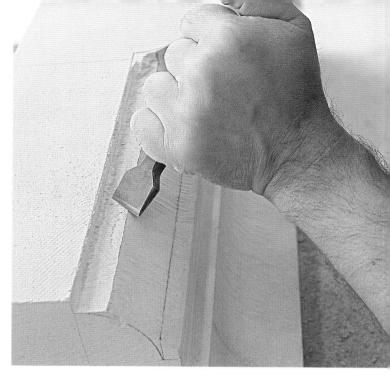

Working a fillet. Note the waste material in the corner of the fillet.

Rounding the surface, by drifting the boaster across the length of the mould, lightly striking with short fast taps.

Chamfer on ovolo, prior to cutting in the fillets.

the waste slip at each end to get the levels. Next relieve the fillet arris along its length by chopping sharply down from the scribe line, then shift to the chamfer side and work from the chamfer towards this (you will need to figure out which is going to be the most effective size chisel to use, as too small could take ages whilst too big may hamper accuracy) until you have formed a ledge from the fillet to the chamfer. Tidy this up, working off all highs, and with a boaster drifted gently in form the internal corner, all the while checking the flat side for true with a straight edge, as well as the fillet shape and depth with the sliding square.

Chamfers

The next chamfers to be cut in will be across the pyramids left, so this time mark a tangent through these at one end; then the point where these intersect the existing chamfer must be applied on exactly the same position at the other end. There are a couple of ways to accomplish this. One is to set the sinking square at the exact depth of intersection and transfer this to the other end. The other is by placing a straight edge on the point; keep it there and rotate it until it is parallel with the fillet edge, then scribe along. This is another crucial point (as if you need reminding) because getting these out of line will cause the curved surface to distort.

Now two more triangular sections are sat waiting to be removed as before, though you may dispense with the claw

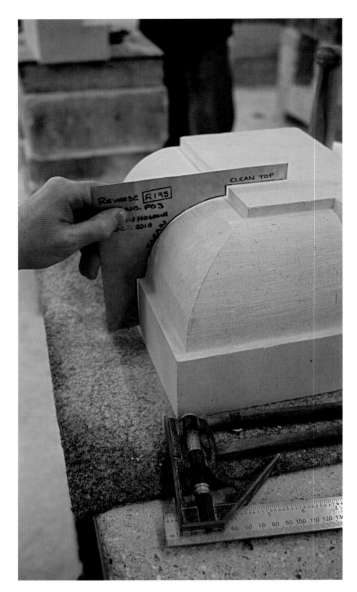

Checking with reverse. The adjustable square is set to measure depth of fillet.

Fillets

Draw a line to the chamfer at right angles from all four points of intersection of the quadrant and fillets, then join them by a scribed line along the face of the chamfer and you will have two slips of stone marked up; these are to be removed to form the fillets.

No pitching here: take a small chisel and cut out a piece of

MAKING A REVERSE

Back to the setting out table for a minute to make a reverse, which is a concave templet used to check that the curve you are cutting is correct (convex reverses are also used but not for this exercise).

These are best made from zinc (as they need to be applied on edge without buckling or getting torn) by cutting through with dividers as before, cleaning up edges with a round file and abrasive paper.

Make the curved edge about two thirds of the ovolo section with enough meat behind to put on the description. Also make a hole for hanging safely – leaving this around will get it bent, and when flattening it back out with a hammer the shape can become distorted.

and just use chisels to cut these chamfers, which once done will give four (flatter) triangular sections that are small enough to work along from one end to the other by using a chisel that is able to cut the whole width in one pass.

There are now eight little triangular sections waiting to be marked up and removed before the rounding of the ovolo takes place, so once they have been worked down you will have a faceted surface close approaching the finished curve.

Curved Surface

Now cut a curved draft at each end, following the ovolo section with liberal use of a sliding square to get them at right angles to the side on which the section was marked up.

Use the reverse to check the small drafts already made to familiarize yourself with the way the curve should look, then using mallet with boaster, standing square on to the worked side, start to work over the edges of the chamfers towards the fillet, from near the middle of the quadrant, very gently by applying small almost tapping strikes to drift over the curve altering the chisel attack angle by rotating the wrist. The trick is to only cut the high spots at the edge of the chamfer first, then gradually spreading the area of chiseling from centre of chamfer to centre of chamfer until there are no more high points – just a curve!

Work a section the width of the boaster at a time to remove the chamfers, but don't worry about getting too exact straight away, as when you can apply a straight edge from end to end then the surface can be worked down as a whole.

Some will prefer to do this stage with a dummy or small hammer to get a constant tapping strike – have a go and see which you feel most comfortable, and more importantly, in control with.

During the constant repetition of this stage of chiselling a curve you will have time to ponder on why abrading the stone by file or sandpaper – as surely the most obvious of solutions to get a curved surface – is not used. The answer will become apparent as we progress. As a practice it is possible and certainly widespread. You will come across some stones obviously soft enough to finish curves with coxcombs, whilst others will need tungsten or diamond grit to remove material. The real test comes when cutting internal mitres and other detail where only a chisel can be effective. Finished stone needs to have a consistent surface, so if you cannot do the same finish everywhere then do not do it anywhere. Realisti-

cally, rubbing out a surface is a good technique when applied in moderation. Before you get to that point, however, you should be competent to carry out all the work by stonemasonry.

During this internal discourse the curved surface, constantly checked with the reverse, will be worked to a fine finish with all the (visible) tool marks running along the curve adding texture and drawing the eye along – stand back, be satisfied with your work and imagine great runs of this at high level in a stately building; polished unnaturally smooth it would resemble some giant piped cake decoration – by your craft and skill it has the hewn character to fulfil its role and be visible as stonemasonry.

RETURNS

The next stage is to return the moulding back along the sides of the stone, so that one side will end in a stop (where the moulding emerges, or enters, a flat surface that approximates to the wall level) and the other will be a return and break (to go back to and then turn to run parallel with the wall face and the first moulding you have made, being cut through the section at the vertical joint of the stone block).

The issue of where the joint in stone moulding or similar should be cut is simple: always cut at, or as near as possible to, a right angle from the face. Never have a stone joint cut on a mitre of any sort – that is for woodwork only.

Marking and Cutting a Mitre

To start the intersection of the face moulding and any return, a mitre will need to be marked on the stone using the mitre-board; as yet there is only cut stone at the face, so a line must be drawn across this to ensure the correct levels are cut to for the return.

Obviously the section drawn on the face for this has been worked off, but the lines from the bedmould are in place and will indicate where the fillet returns and then carries through to the back of the stone where the templet was applied and marked around. The moulding must end at the stop (the wall line) so where the stop is marked on the bedmould drop a vertical down to reach the bottom fillet, and scribe this line in. Now that the limits of the working are defined it is time to mark on the mitre.

Mitre

Place a square on top of the stone using the handle as the base and the blade upright near the mitre area; this will be used to ensure the mitre-board is vertical. Move to the corner and place the mitre-board flat against the square's blade and hold it in position whilst you manoeuvre the board to cross from the top fillet point of return down to the bottom fillet point of return (the corner of the stone). This being a fiddly operation until you are used to it, it helps to rest the end of the board against your body for support (tucked into the hip). When the board is sloping up the corner of the stone held in place on the square and touching the stone at the top fillet and either on the ovolo or the bottom corner, place a pencil against the board and manoeuvre until the pencil (kept flat against the board) can touch the top corner and the bottom corner just by moving along the mitre-board (which will be held steady); use a pencil that is long, sharp and preferably of polygonal section. Then keeping the flat on the board trace a line on the stone from top to bottom – this line is the mitre line that the runs of moulding will intersect at.

For interest hold the templet level with the outside fillet line to the corner and glance across it; when looked at from a right angle to the face and towards the stop it will match the mitre line.

Now comes a bit of practical geometry that will confirm the

Stance to mark mitre. The mitre-board (a strip of hardwood true and flat) is held against the square at the top and tucked into the hip at the bottom. The pencil is flat against the board and the whole is manoeuvred so the point of the pencil can touch the top and bottom points of the return without moving the board; then the line is drawn from these points across the mould as the pencil follows the contours.

exactness of the mitre line and give the next stage to work: on the scribed-in section at the back of the stone mark a chamfer that crosses the curve as done before; where this line intercepts the block draw along parallel to the fillets on the top and side. These lines intercept the section, and a line drawn between them along the mitre board held sideways on will give the developed form of the chamfer. This will touch on the curve of the mitre line exactly; adjust until you get it right – and then do it again to make sure.

Another Chamfer

This chamfer needs to end at the stop by cutting three relieving drafts, two along the chamfer and one at the stop. Then working into the mass of the stone to be removed at the stop end, take out enough stone to allow a draft that joins up the chamfer limits and its other face starts to make the flat surface of the stop (continuous with the wall line). At the other end cut the draft across the curve and then as before work the stone down to the chamfer

Fillets

The same methods are used here as cutting out the ones on the face except where they stop, and this must have the arris relieved before and during work. The important part is to mark everything out correctly and, as always, work accurately, making sure that the flat surface of the stop/wall line is cut in sharp from the beginning. It is wasted effort to leave this a little proud to allow working in later; dressing into the stop is fiddly enough without having to keep going over it – so aim to get this right from the start. Working to tight tolerances is beneficial to confidence and it will help in the long run.

FINISHING

Once the fillets are in you will have the two points running along the curve. These have to be chamfered by starting off by eye without the aid of the section marking. Place an inch chisel on the mitre at a tangent to the drawn curve to cut the start of a chamfer and lightly 'pitch' it upwards – this is going to be the tangential line of the next chamfer, and by taking parallel lines from the points of intersection back to the stop you can mark out the next block of material to be removed as before. Repeat this for the usual eight chamfers (or more if required). Then using the boaster take off the high spots, checking all the time with the reverse held at right angles to the run of moulding, paying particular care to the way the mitre is worked in with the emphasis of the strike cutting across and into the stone. As the curve is trued up the mitre will resolve itself, being sharp on the furthest limits of the curve and blending into itself near the fillets. At the stop, trim in the corner where the ovolo enters the wall where, with all levels correct, this will also resolve itself.

After working this, the difficulty in using abrasive methods to form the curves should become apparent, as it would be almost impossible to work into the corners neatly.

It can be useful on certain stones to start the first chamfer in by sawing across the wall line (or disc-cutting); in practical terms this can occasionally be useful, but hand-working is often just as quick – think about the reasons why this should be so.

Return and break, showing the stone positioned to work the two chamfers into the internal mitre. At this stage the mitre-board is no use, and all must be done by eye and with edges.

Return and stop.

Return and Break

To accomplish this involves straightforward application of the techniques learnt so far plus a few added processes, the principal of which is that whilst the external mitre can be marked up on the stone, the internal mitre can only be resolved by working the intersecting mouldings in concurrently.

So lining out the fillets from the section on the side of the stone through to the intersection with the return from the face, draw in the limits of the first chamfers; now work these

down as before, noting that they will meet in the corner. Work carefully here until you are satisfied that the two chamfers meet in a sharp internal mitre.

The fillets are formed in the same way, remembering that the crucial point is they meet in the corner and all the lines (of the corner) run in a straight line from the top of the bed fillet to the bottom of the outer fillet, checked by eye.

Everything here has been explained previously; the main divergence is that the curve is formed by boasting over while gently drifting in sideways to follow the line of the (internal) mitre – as you get into the corner it will be harder to use the reverse so ensure you work a quality curve, checking from the finished curve by straight edge.

Finished

The above explanation would appear to get less wordy the more complicated the manoeuvre undertaken, but consider how there are not that many different actions involved in this exercise, the main issues being the setting out of mitres and the working of the first chamfers. Be critical of your levels of expertise as well as the quality of the work, and do not rush things – accuracy is always going to be mentioned and stressed; measure three times, set out, mark up and cut to the line is the credo we work to.

CHAPTER 9

ARCH AND VOUSSOIR

Nowhere do we see the work of the stonemason displayed to such quality than when contemplating the arches that allow soaring gothic vaulting to hold their majestic spaces or that support chasm-spanning aqueducts striding across mountain valleys.

Arguably one of the greatest advances in building construction was the invention or to be more accurate the development of the arch. With this, man was able to look forward to passing through wide portals into a world of large uncluttered well-lit interior spaces, whilst the knowledge of vectors which the curve engendered has underpinned much of our technology. The widespread adaption of this engineering feat was really the point at which humans left behind their reliance on the materials for buildings being used as found. More specifically it completely changed the limits of designing for such materials.

Classical (Greek) and Egyptian architecture, whilst impressive, was hampered by the inability of stone to support a roof or weight above gaps with walls further apart than the available length of stone, resulting in buildings studded with supporting columns, massive walls and relatively little openings – big, grand but very cluttered.

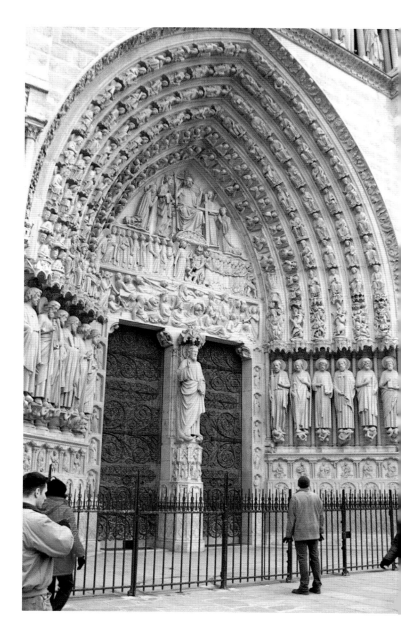

RIGHT: Notre Dame, Paris, a supreme example of the show business approach to illustrating the front of a medieval cathedral.

OPPOSITE: While the ancient structure falls into disarray, the arch continues to do its job; built properly, with no outside influence, the arch appears immortal. Ancient masonry structures are incredibly intriguing to the chroniclers of history and the writer of the romantic novel.

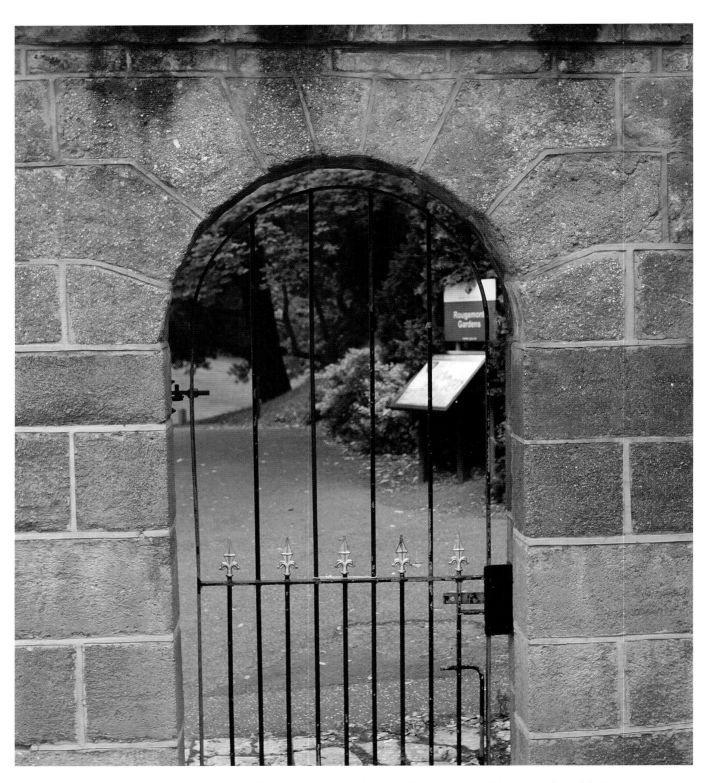

Simple gateway, Exeter, using different stone to add beauty. The pointing, however, is modern cementitious and ugly – original stonework was always about the stone, not the mortar. The relatively modern practice of highlighting the joints of masonry has produced some woeful sights.

*Ambulatory,
Coutance.
This part of a
cathedral was
designed so that
pilgrims could
walk through
the church (with
horses, dogs
and so on) and
not disturb
the services
which went
on continually
throughout
the day during
medieval
times. The
large clustered
columns, a hint
of earlier style,
jar slightly with
the elegant
vaulting.*

WHAT IS AN ARCH?

Put simply, an arch is a means of moving force away from the centre of a gap into the walls either side, which enables the structure above to stay where you want it when the material is removed from beneath. While there must have been a eureka moment at some point, with all successful stonemasonry the technique relies on physics and geometry. Nevertheless it may be interesting to consider how this miracle could have occurred.

As spans across openings increased there may have been failures where the lintel would break and possibly fall out. The result would be a roughly curved space edged with stones pinned down by the weight of the masonry above. The astute stonemason would note how this occurred and consider that by gradually stepping out stones from the side you could reach out further across the gap. This led to the development known as corbelling which, whilst effective, still had its limitations: as the gap widened, more weight was needed to counterbalance the projecting stones.

Simple and playful demonstration of how stones in an arch support each other. Find two abutments, pile sand up between centring and place flattish stones packed in a curve from one to the other; remove the sand and the arch will take up its position as all the stones lock together.

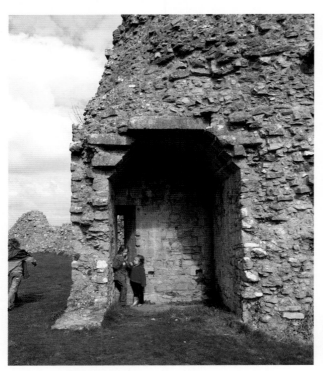

ABOVE: Corbelled vault, Corfe Castle. An early (possible) arch development relying on stones stepping out, the weight of the wall on their tied ends levering them up to counter the forces.

RIGHT: Simple two-stone arch, Exeter. It is easy to see the development from two leaning stones, shown in the picture on page 116, to curved soffit.

Now imagine one day a shift of weight caused the corbelled stones to tip inwards. They wedged against each other where, by locking into place, they utilized the laws of physics to support themselves and span the gap. By replicating this situation and refining, using the wonderful skills of stonemasonry and dressed stone – eureka, the arch was created!

An arch is the embodiment of teamwork as it relies on all its members performing their task correctly; without each one contributing equally it will fail.

Stonehenge lintel construction; note the tenon sticking out of the top of an upright to locate the lintel. These are called Sarsen stones, derived from the term Saracen meaning stranger, as they are unusual for the local geology. This is probably not the original configuration, as the majority had fallen down when the whole was subjected to a major reconstruction in the early twentieth century.

DEVELOPMENT OF THE ARCH

As stonemasonry developed from the crude lintel to semi-circular arched barrel vaulting (which in truth is a ceiling composed of lots of arches, face to back) there were still problems about weight and span. The other issue was that all rooms made this way, if they wanted to meet with ceiling at the same level, had to be of the same height and width, with the proportionally massive construction of the walls providing the necessary mass to restrain the forces threatening to push out the walls. Arches like this were regularly used across the Middle East, from well before the birth of Christ.

The first use of a pointed arch was for drains at Khorsabad c720BC, but this technique was not widespread during the Dark Ages, and as the semi-circular arch had its limitations, something was needed to overcome this.

Without going into the full details here, it was in France around the start of the second millennium that an enterprising stonemason created a method of spanning that could be taller than it was wide. The lancet arch kick-started the great surge in fantastic building known as Gothic (a derogatory term coined by Palladio) that defined the era we know as the Middle Ages, when stonemasonry was probably the most important occupation in the secular world.

The pointed arch not only made for lovely looking spaces; it allowed buildings to be taller and lighter. This was because the forces that were pressing down on the walls could be directed and contained into relatively small cross-sectional columns and walls; the advantage here was that the building would not fall down just because it was constructed beyond the limits of contemporary knowledge.

LEFT: A buttress in the wall of Wells Cathedral, seen here as curving stones set into the masonry. These are arches designed to push forces down through the mass of the wall.

RIGHT: Salisbury Cathedral nave, with it soaring lancet vaulting, shows the lightness of structure over an immense space.

Leaning stone arches, Exeter.

TOP LEFT: *A clerestory window, Coutances Cathedral. The holes in the wall are for the putlogs of Mediaeval scaffolding.*

TOP RIGHT: *A Mediaeval crane recreated in the grounds of the Abbaye de Hambaye in Normandy. This would have been built in the roof of a Cathedral to haul the massive stones up.*

MIDDLE RIGHT: *Romanesque arches remaining in later gothic cathedral, Coutance. Rather than tear buildings down, many great buildings would incorporate earlier masonry as part of the structure.*

BOTTOM RIGHT: *Lancet arches, West Front of Salisbury Cathedral. Behind the row of statues, holes can be discerned in the wall, behind which is a gallery where a choir would sing; the effect of fully painted statues with heavenly voices would have been enough to convince anybody of the power and glory of God and the church.*

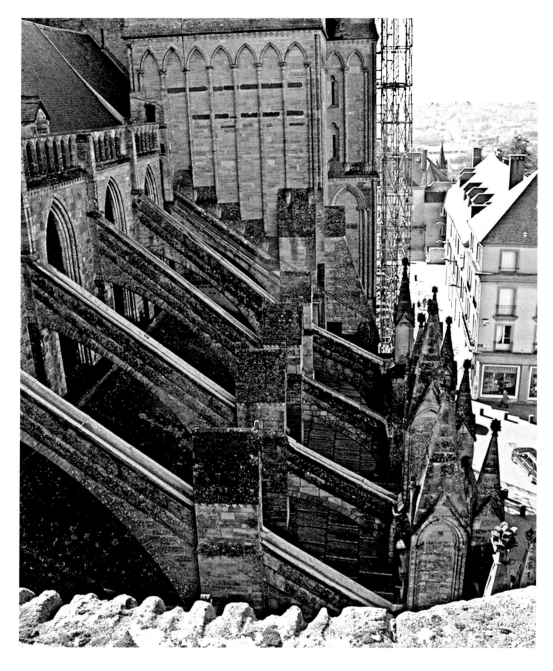

Flying buttresses, Coutance, the arched soffit allowing forces to be transferred down and away from the walls; the pinnacles at the right are to provide downward force to help the stability. The buttress was often not tied into the masonry at the wall of the church; rather it rested against the surface (with a bed of not too hard mortar) and could slide on the wall, preventing stiffness and rigidity.

Flying buttresses absorbed the forces coming out of the building and, by the placement of heavy pinnacles, channelled them down to the ground – effectively pinning these churches to the ground!

One of the most glorious uses of the arch is in Wells Cathedral where a vast swooping armature of stone was inserted into the crossing, to support the great weight above after it was found to be in danger of collapse.

MAKING A VOUSSOIR

What we are going to do here is to make a voussoir, which is a single stone for a simple Romanesque arch (known in Great Britain as a Norman arch, being introduced after the conquest). Obviously making one player for such a team event is pure exercise, so if it is possible to find other aspiring crafts people to contribute a piece to the whole you could fully

Wishbone, crossed arches inserted in Wells Cathedral to take massive forces safely down into the floor, preventing the toppling of the masonry above.

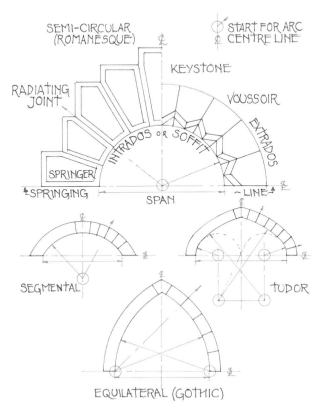

SEMI-CIRCULAR (ROMANESQUE)

START FOR ARC CENTRE LINE

KEYSTONE

RADIATING JOINT

VOUSSOIR

INTRADOS OR SOFFIT

EXTRADOS

SPRINGER

SPRINGING LINE

SPAN

SEGMENTAL

TUDOR

EQUILATERAL (GOTHIC)

appreciate the later exercise of fixing the arch in place. Or perhaps you may aspire to do the whole thing yourself – bon courage!

The joy of the (Romanesque) arch as proposed here is that it can be composed of identical pieces so you will only need one set of templets for the whole structure.

Once the span has been decided the distance from the exact centre to the edge of the opening is the measurement to get the curve of the intrados; this point will also give the origin of the extrados radius and any concentric detailing.

Bedding

Here is where the direction of the natural bed of the stone becomes pertinent, as in most applications the best way to lay stone is such that the compressive forces should always be at right angles to the bed, effectively squeezing the stone together. At the springing line this will be horizontal with the rest of the wall masonry, rotating to become vertical at the keystone of the arch; the notion here is of radiating beds running from the centre of the span through the intrados to the extrados.

Components and terms of an arch.

VOUSSOIR

This shows the section of the voussoir with a roll mould.

To develop the roll for the joint draw a circle below the section and divide the perpendicular axis into as many parts as appropriate (seven here) and take their heights across to the circumference.

Where they cross take this line up to the section.

Now divide the joint line on the face into the same number of parts. (This can be done by drawing and marking another circle on the mould at right angles to the centre line and marking as before – don't forget the bottom half as well here, always numbering from the centre).

Bring these across horizontally and where lines with the same number cross, use these nodes to form a smooth curve that will give the (elliptical) developed section.

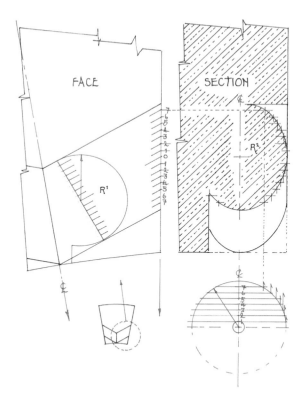

This shows the section of the voussoir with a roll mould.

Templets

To get the external dimensions of the voussoirs, you have the stone length as the measurement from the cusps of the intrados to a tangent on the extrados; the height will be distance between each corner on the extrados; the depth will also be as required for the wall. Thus it is cut from a squared block ending up as a wedge, with a concave end at the narrowest and convex at the widest.

Here is where the size is governed by what you can obtain in your particular circumstances, so using common sense divide the semicircle into equal pieces that will accommodate the stone used. Don't attempt another Arc de Triomphe – keep it practical!

DIVIDING A LINE

It is relatively easy to divide a straight line into equal lengths by using calculation, but it becomes difficult to apply mathematics when the line is curved. Fortunately, as artisans would not have had ready access to the maths for this, there developed a practical method that is capable of great accuracy.

Example

If an arch needs seven stones, then using dividers, set to a guesstimate length of a seventh of the distance, step off spaces along the arc. Now you will have either gone over the length or come up short; so look at the distance of error and adjust the dividers by approximately one seventh of this, in or out as appropriate.

Step off along the arc again and you will notice how much closer you are to the exact measurement. With the slightest adjustment it will be possible in the next stepping off to be right on the mark needed.

This method can be used for any setting out – the speed in getting accurate will come as your spatial awareness develops.

On a bench gaffer-tape a small square of zinc that has a nail hole tapped in the middle (this is the arch centre). Then tape a piece that is exactly the size of block of the stone to be used so that it laps both the intrados and extrados distance from the centre, the farthest edge positioned tangential to the extrados.

Voussoir with roll in progress.

A trammel bar is used for scribing the long arcs – you can make this by using a length of timber with holes drilled on the three points and sharp nails driven through; set the point in the centre piece of zinc and describe the arcs onto the templet zinc.

Lines scribed to the centre from where the extrados meets the long sides of the metal will give the centre of beds. Remember here that these stones, as all stones used in masonry, are not a contact fit and require a suitable gap between them for the mortar – my personal preference is for 3–4mm joints. So the templet must be marked and cut to lines inside the marked ones – parallel to the centre bed line, not as radiating lines or the joints will become wedge shaped. Where they cross the inside arc will give the intrados.

With the waste cut off and the edges cleaned up you now have the templet for all the voussoirs. Mark it up and drill a hole so it can be hung on a nail.

ONTO THE STONE

Place the templet onto the stone and line it up so the corners of the intrados touch the short side and the extrados corners touch on the long sides. Scribe around it and then transfer the templet to the other side; do not turn the templet over as it is essential to keep any discrepancies matched. The two should line up with lines that are right angles to the face.

The two wedges of waste stone can be removed in the same manner as the flat surface. However, as the surfaces are to be inside a joint that is crucial to the structural integrity of the arch, a dead smooth surface will be counterproductive. So pitch off on the line, put a narrow draft (remembering to cut into the corners to prevent spalling) around the edge, and leave the centre level with the drafts and with a textured finish from a claw or chisel furrows that will give the mortar some grip on the stone – this is known as keying the surface. Do not make the mistake of hollowing out this surface, thinking it will aid fixing; this can create pressure points on the arris and also makes wedging difficult.

The extrados is the next task – the procession here as with most stonework is to complete first the parts that have least possibility of getting damaged during working.

Extrados

It is straightforward work to cut curved drafts (that you can check by a square from the face) along the scribed lines and

Sand filled inner tubes to support work at various angles.

At the straight ends use a boaster to work the draft in at right angles allowing you to start into the curve; use the straightedge and square from face to back to stay accurate. With this method you can use the drafts as guides and work the stone by boasting the curve into the centre, then working the same from the opposite edge. This surface is going to remain uncovered and so needs to be worked neatly by keeping all the tool marks neat, tight and in the same direction; this good even texture will give the soffit character.

Before we go any further it will have become obvious that without getting the stone into a suitable position it is difficult to work the myriad surfaces efficiently, so by now you should be adept at securely wedging a stone of whatever shape into a position to work it. Try getting hold of some old-fashioned sand bags to help. You can also make supports from lengths of inner tube, filled with sand (not too full as it needs to be flexible) and tied at each end; these have the advantage that they can be any size and do not hold dust.

Joggle

To help in fixing stones that have a structural role to play in place, it is acceptable to cut a joggle in the beds (here exiting at the extrados) to be filled with grout.

remove the centre waste, accuracy being checked with a straightedge from front to back. If you decide to pitch, use a tool appropriate to the curve and avoid knocking off stone you need to keep.

Once again this is a constructional surface and so should be worked cleanly on a margin to the arris of the face, with a lightly keyed surface for the interior.

Intrados

Firstly do not use a pitcher on the line of this! If you are determined to remove some waste with a pitcher use it well into the mass of the waste, away from the finish lines. As this is a concave surface cutting the drafts will require a bit of care by approaching from an oblique angle that does not cut into the stone to be kept. The angle of attack will be determined by the speed of the curve – just keep checking off the face with your sliding square.

OPTIONS FOR FINISHED STONE

Tooled Margin

One of the joys of working stone is the imparting of different finishes to the surface, which can highlight openings for example, or add emphasis to plinths. The simplest way is to use a boasted finish across the whole face, but to add a more formal dimension finishes are enclosed in stones by a margin. This is effectively framing the stone and allowing anything to happen in the centre because the margin provides sharp edges to build to. Thus the centre can be finished in a distinctive way – popular treatment in classical and baroque architecture.

Here we will add a simple finish to the edge by creating a tooled margin to the face of the voussoir, after which you may wish to finish the centre of the stone with one of the many effects available.

You may want to have a practice on a spare bit of stone to

Setting out tooled margin.

the length of the margins you may have to slightly alter the distance between the last few lines to take up the slack; subtly done and with the optical puzzle of the corner this discrepancy can become unnoticeable.

Forgetting the corners for now, place the chisel vertically on the first complete line and with a small metal dummy or hammer give the chisel a set of strikes that whilst rotating it to the horizontal will cause it to cut out a trough that fits in between the marked lines. The exact sequence and method is bespoke for the stone in use and will rely on your interpretation of what is needed and suitable. Once the first has been done place the chisel on the next line and repeat the process exactly, and so on until you have filled the margin with perfect fluting of regular spaced sharp peaks. On the curved margins you will have to apply a little more pressure to the chisel at the wider end, giving it emphasis to make the taper, and at the corners move the chisel out so that the flutes intersect along the diagonal; it will need a bit of fiddly work, in the reverse direction, on the very corner to prevent spalling.

Running Mould

The above instruction will give a simple arch. Another option though is to apply a decorative moulding to the curve that will give you practice and bring vivacity to the arch, as shown on the templet drawing.

Here we will attempt a chevron style, of a roll mould, simple but very effective for our needs.

Basically this is two roll moulds that angle down to intersect in the centre of the stone. Where they exit at the joint becomes, because of the angle, a developed section of the roll, which as the enlargement is only in one direction becomes an ellipse.

Apply the templet to joints and scribe in. Scribe lines from the outer edges of the roll up to the face; take these down to the intrados to intersect at a centre line drawn on the face of the stone.

Now relieve the arrises along the triangles marked out and down to the new level of the face at the intrados. Using the standard technique cut drafts and take out the waste using a claw, then boasting down to a fine surface and ending up with a squared-off ledge.

The section we are using is very similar to the style used on Romanesque arches throughout Europe, probably because as a journeyman our medieval stonemason would be limited to

get the hang of this technique, though at this point your skills should be up to carrying this out competently.

Decide on a width of margin – they are usually about 25mm wide, perfectly suited to an inch chisel – and set the sliding square and scribe around the face of the stone to this measurement. Then divide the margin by drawing lines along with the square at 3mm/$^1/_8$ in spacing. This is to allow you to place your chisel in the right place to cut fine flutes; on the intrados do the spacing steps on the edge of the stone, whilst the extrados should be marked on the inner, scribed line of the margin. The flutes on the corners will step down in length meeting along a nominal diagonal.

As it is improbable that the lines drawn will fit perfectly into carrying out work with the equipment he could carry from town to town. This means he would own a collection of templets, which he would have probably manufactured during his training, using setting out equipment held at the lodge; and these could be used for a variety of jobs. This templet could be used for plinths, stringcourses and if added to the templet of the quatrefoil would be good for deeper doorjambs and surrounds.

The evidence we have to confirm this is the arches that have various sized voussoirs, which hold unequally spaced chevron moulding. The stonemason would be working on the construction of a doorway, and once he knew the size of the opening he would go to the quarry and on a flat surface describe the arch using cord and a scriber. The quarryman would be instructed to supply stones to create this. He would use whatever he had to hand. So as long as it fit the intrados to extrados he did not care how high (the measure from radiating joint to radiating joint) it was – why should he waste time cutting down stone when he is fulfilling the order of an arch? When these stones are handed to our hero, rather than going through the process of setting out the chevrons equally around the arch – an arduous task which would end up with a multitude of developed sections to work out as the

joint intersects the roll in different places – instead he places the templet on each joint surface and then gets the roll to intersect in the centre of each stone, creating the arches shown.

Roll Mould

To make the roll is basically an extended ovolo that cuts down into the stone at the fillet; the difference is that the section is an ellipse. On the setting out for the section, mark tangential chamfers on the circle and bring them up to the elliptical section. You will now have two points of reference: one is the angle of the developed chamfer that can be transferred to the stone; the other will be the distances of the chamfer limits on a box containing the circle. Use these to confirm each other's correctness.

Cut the chamfers as other curved sections, taking care at the intersection to preserve the mitre; the depth can be checked with a sinking square or by cutting small triangular reverses and using them to monitor accuracy. Then work over with the boaster using a reverse to check accuracy, keeping the tooling even and parallel to the roll.

Arch with unequal voussoirs, Piddletrenthide, Dorset.

Victorian centring, Woodchester Mansion. Note tas de charge stones below the centring, for the subsequent vaulting to spring from; these complicated looking stones are in effect a rayed springer stone.

CONSTRUCTING THE ARCH

Centring

Here we have a gap in a wall of the right size waiting to be spanned by the arch you are going to construct. You have the stones and the mortar. What next?

The stones must be safely assembled together, and to do this we use centring, or formwork – a frame that allows all the stones to be placed correctly and supported.

Centring is of wood and should be an example of respectable carpentry skills, as it needs to be accurate, strong and durable enough to cope with life on a building site. It is best made with slats as the support so that wedges and levers can be used to level up the voussoirs.

Set the centring in the gap resting on folding wedges at the jambs all supported on a solid construction of suitable height, preferably an integral part of the scaffolding you will need to work from. Make sure that all the stone and materials are sat on the scaffold as well as your comprehensive toolkit, strips of lead, softening, packers (slate or plastic) and wedges.

Common practice is to have previously laid the arch out on the ground to determine the size of the joints and see how the lines run. From this you will be able to use packers of an appropriate thickness when erecting it dry (without mortar) as we are about to do.

Dry Run

Place packers on the springing line bed either side of the gap and position the first two voussoirs on these. Proceed then

Model of centring. Well made, it is a good example of quality site joinery.

Laying out a lancet arch to check before building; note the carpets to prevent damage to arrises.

placing packers/stone/packers until all the stones are sitting in the right position on the centring.

Now here is where personal judgment and common sense comes into play. If the centring was dropped out now, the stones would all shift and lock into position separated by the packers, and the chances are that it will not be as perfect as you hoped with uneven joints, dropped stones, unsmooth curves and possibly leaning out of true. So set the centring

back in position and knock in the folding wedges to lift and support the voussoirs; note the positions and any other issues and consider how to set this right next time before dismantling it prior to doing it right.

When all is in good order, take note of any singularities, such as joint width and thickness of packers needed, and write this in pencil on the formwork. Give each stone a number on the extrados joint to prevent mixing up stones.

The reason the packers should be of slate or plastic is so that the stones can be manoeuvred on them without jamming up; also these will not compress. They are placed in the joints back from the faces, evenly around so the stone will not rock. (Hollow joints here would be a hindrance.)

FINAL FIX

When you come to actually build the arch, set the centring slightly high by just tapping in the wedges a touch. This will allow the stones to squeeze into place when it, invariably, settles as the centring is dropped. So lower the centring to let the stones grip and then carefully set and lock them all into their exact position by soft-nose levering and tapping in thin wedges. Once the joints are all even, curve correct and level

Waiting for keystone, showing a joggle. (Photo: Ian Constantinides Ltd)

Keystone lowered into place by split pin lewis. Small wedges are visible; also softening is placed in the joint to prevent pinching in this fine jointed work. (Photo: Ian Constantinides Ltd)

with the wall, gently tap in slivers of slate to lock it together. Remember any push to one area may dislodge elsewhere so be careful. Also ensure any permanent wedges and packers are well back from the joint so they can be covered by mortar.

Pointing Up

Now dampen up the stone, especially the face, and start packing a very stiff mortar into the joints. Use a slicing motion with the trowel edge to move the mortar into the space, always keeping the mortar off the face and sponging any mess off immediately.

Once there is mortar completely packed into the joints around the face, soffit and back, a sufficiently runny grout can be poured into the joggles to provide a complete fill. When work is done, cover with hessian and polythene to keep humid while the mortar goes off, lightly spraying when necessary to prevent it drying out too quickly.

The arch is now part of the structure and will eventually be able to support great weight. But what must be remembered is that it will fail if the force downwards is enough to push the stones outwards along the wall line, because the only restraint of this is the grip of the mortar on the springing line. So to fulfil its promise the arch should be bounded by the rest of the wall built up around it. Then you can do what every stonemason wants to do on their newly made arch – stand on it!

Romanesque arch, Sherborne Abbey.

CHAPTER 10

THE QUATREFOIL

What the layman usually comments on when looking at a gothic cathedral is the movement and energy of the tracery design in vaulting and windows, commonly attributing to it an organic genesis with similarities to growing things; this is a misguided perception, for the true ideal of gothic is control and precision. Take time to study a tracery window and with a sketchpad work out where the curves start, and you will become aware of the beauty and simplicity of the geometrical setting out, where judicious placing of a couple of points allow arcs to spring, touch and weave a poem in stone.

SETTING OUT

With a square piece of stone of the correct depth, we are going to produce a corner of a frame containing a tracery quatrefoil, which like the voussoir of our Romanesque arch is a single design that can be repeated to make a complete piece; the section is also useful for many other applications.

Medieval vs Victorian

There are two methods that will decide the approach. The conventional approach is to design the quatrefoil (or rather a quarter of it) and set this down on the facemould. Obviously the arcs and lines denoting the run of the mouldings will intersect at the joint and from these the section templet can be made; this relies on accuracy at all stages, because if the lines are out the templet will not match up with the joint on the other side. Done correctly this

RIGHT: *West front, Exeter Cathedral – brainwashing, imposing and educational all in one go!*

OPPOSITE: *A section of tracery, with its curves and texture, awaits the chance to become the epitome of our fascination with geometry in structure. No mere accident gives us the decoration of stonework; it is practical mathematics writ large.*

The four stones to make a complete quatrefoil (in this exercise we use the term quatrefoil for each stone as well).

will give a crisp and orderly stone that could be termed the Victorian style.

As connoisseurs of medieval stonework will point out this Victorian style does not quite fit with true gothic working practice. Our journeyman stonemason, having served his time on one of the great European cathedrals, took up his equipment and set out on his own, visiting a provincial town where he regaled local parishioners with his sales spiel and was subsequently contracted to make an agreeable feature for the church – a lozenge shaped quatrefoil made of four stones. In his toolkit he has his own templet cut from a piece of oak (with appropriate reverses) of a section that impresses the locals (it is also cheaper than designing a specific one for this job). He applies the templet correctly to the joints of the four pieces of stone supplied from the local quarry. As he is hoping to get it done quickly (more profit) he does not complain if the stones are slightly out of true. Rather he marks the face in situ on the stone, juggling the centres to make everything meet at the joints, knowing the finished stone will be enough to impress the patrons who, unaware of the slight discrepancies, will happily reward his endeavours.

This (medieval) style of work with its almost freeform setting out, which often results in idiosyncratic deviations, horrified the Victorians, who were interested in the history of 'picturesque' building; they were not the master masons of yore but aesthetes without practical skills. So the craftsmen of

that period would be rigidly trained and utilized without any freedom of expression (except for the chance to add faces to label-stops or similar). It is an illuminating exercise to walk around a church and spot their repair approach; you can tell by the precision and quality of the setting out.

Obviously our stone will be square so this problem will not occur, but we will work to the section, and the facemould will be used as normal (which is more for reference), allowing you to determine which is more effective.

MARKING UP THE STONE

Here the templet of the section is set out and the lines that intersect the face of the stone are divided up to carry around the edge and to arc around as the foil to the meeting point at the cusp.

With the moulding set out on the templet, place it on the joint as normal. Scribe it in, then rotate it around the corner

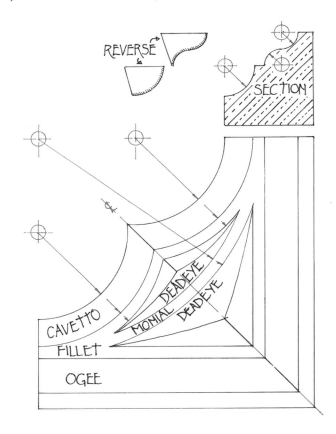

Drawing for setting out.

of the stone and repeat on the other joint; this shows the narrowest section of the stone at the farthest edge of each foil.

Now we get down to some geometry, which could be set out as a facemould, with the advantage of not having to do this for each stone, but realistically you could know where all the points go from a good quality sketch, and in truth you do need to do this for each level. Bisect the stone from outer corner of the frame to the nominal centre of the lozenge and trammel around the outer two edges to define the part of the section that forms the frame, in this case the ogee moulding. The flat part of the moulding, which splits from the section that follows to the corner of the frame and goes into a curve, is the monial.

Next bisect each side at the monial level to get the centres for the foils and describe the arcs of all the levels from these points. Where they intersect on the main bisector is where to form the cusp. Be sure to carry all lines around the stone to find the matching points on the back as well; this is necessary for subsequent construction when the stone is worked down to new level, as the setting out scribing information (facemould) will be lost.

CHANGING THE LEVEL

The frame of the piece should be cut as a separate exercise to the monials, which are worked at a lower level. First a sunken face must be made by squaring off the frame moulding down

Boasting the sunken face.

to the fillet and scribing this height along the open edges (setting the face of the foil's depth).

Cut a hole into the corner of the foil area, at the intersection of the frame to get the working level. Chop relieving cuts on the other two sides of this square inwards from the frame, and keeping to this limit sink a draft down to the level, checking the depth all the time with sinking square. Once the levels are in, pitch off to the line and, depending on the amount of stone left, punch or claw down to a couple of millimetres above the surface; then boaster across, keeping the tooling very fine as the surface will need to be scribed on.

Work the vertical faces of the frame stone down to meet the sunken face, checking for square at all times, drifting into the corners and at the intersection to end up with a sunken panel bounded on two sides by the unworked frame.

Squaring frame.

FOILS

Now on the new surface bring the lines back over from the back and re-mark the centres; then scribe in the fillet and the monials that describe the foils. In the corner the monials will pull away around the foil, but also carry the width of this fillet up to the corner where it is now bounding a triangular area with a curved hypotenuse – when this is cut out it becomes a deadeye. In circumstances where the curve of the foils is not pronounced, the area of flat surface can be large enough to allow the insertion of another deadeye.

Taking the innermost diameter as a limit, cut, using checks

and drafts, the two circular cutouts that start at the joint and intersect at the cusp. By some deft manoeuvring of the stone and angle of attack, you can rough these out with a bullnose working inwards from the edge, and then finishing off with boaster to give fine tooling perpendicular to the face. Take care to always cut in towards the centre away from delicate edges and down from the cusp.

Foils worked out, the drags are used on this soft stone to good effect.

Roughing out foil.

Chamfer

Marked on the edge of the foil at the section on the joint will be the chamfer, the edges of this should be scribed on the upper surface and the wall of the foil and with deft chisel work these lines can be joined by a worked surface that will take the chamfer around the foil.

The chamfer where it meets on the cusp will need to be worked inwards at both sides to get the correct intersection, which will be a development of the chamfer angle.

Punching out waste.

Cutting ogee on frame and cavetto on foil.

To a Cavetto

Usually this chamfer will then be worked down to a cavetto, which is a good practice and will give more life to the foils, making the whole stone appear lighter; the one here will be segmental, which is shallow compared to elliptical or quadrant, but still effective.

Cut a draft in from the edges of the chamfer, with a half inch chisel, angled into the centre of the chamfer and down to the edge of the cavetto; do not go to the line yet. With the bullnose take out the material left to the line, in a series of curved drafts, making the tooling finer as it comes to finishing, checking all the time with the reverse and remembering to work in from the ends.

The most delicate operation will be the intersection at the cusp, requiring some deft gentling-in work – be careful.

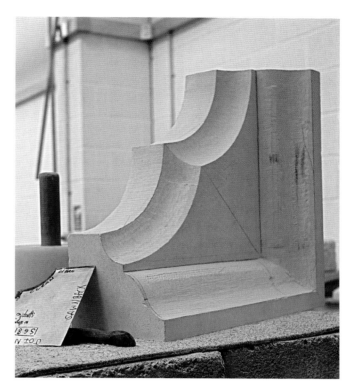

Cavetto and ogee worked.

Working cavetto with bullnose chisel, on a double-sided stone.

THE FRAME

Now here is where good accurate stonemasonry is an asset, as the moulding around the edge is all cut into the blocked-out stone that you have prepared, so check all the surfaces are level and square and do any trimming in now. Do not be tempted to hope that discrepancies will be lost in the curves of the moulding – always start from a good level.

To cut an ogee mould mark a chamfer from the outer edge of the concave section to cut across the convex curve, then carry these lines into the corner from both sides.

Cut relieving drafts along the edges and an exact draft on the chamfer at the section (joint) end. Work the material down with a claw and then bring to perfection with fine boasted tooling. The two chamfers will meet in the corner giving a mitre line that will line up with the centre cusp.

Ogee Mould

Once the two drafts of the whole mould have been cut either end, the first stage of this mould to be worked should be the concave part, constantly checked with the reverse to given levels and using a straight edge to check along the run.

After the (concave) shape has been channelled out by a combination of bullnose, claw and boaster (being careful when drifting in the mitre, not to cut into the other side) at the top of the chamfer the edge of the scribed line will still be evident; if not, rescribe it.

In the scribed line, working along the frame, position a very sharp chisel (from inch to boaster depending on the stone)

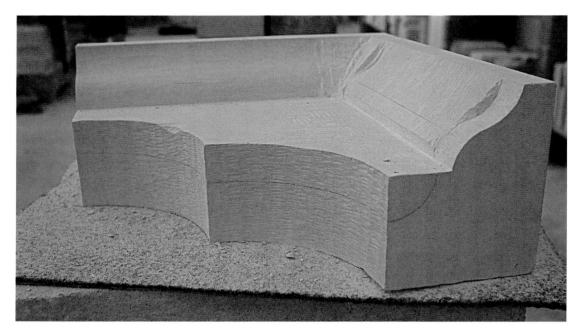

and sharply pitch into the curve; this gives excellent definition to the arris and makes the curve appear fully hollow.

The remaining part of the mould will have to be chamfered down, similar to the ovolo, and with a boaster the curve can be gentled into the concave and cut sharply into the meet of frame and fillet.

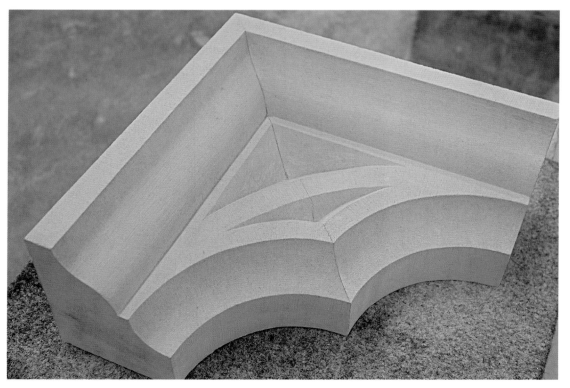

Finished stone.

Deadeyes

When cutting deadeyes it is necessary to go slowly at first to get used to incising, as there is only one line to follow – the other is an imaginary one at the bottom of the cut.

Prior to cutting, draw lines inside the deadeye that are exactly bisecting the angles; the curved ones will best be judged by eye and drawn with a French curve. This is where you are heading with the incision, and at this stage you will be aware that at the points it will be shallow, deepening to the centre; this is the same as in lettercutting and will follow the same process, with some slight difference.

For the one in the corner, from a narrow point chase the line along the centre, to the centre point; turn the chisel to the opposite angle and cut again to make a V of an included angle about 60 degrees, along the centre line. Allow the depth to increase as you approach the centre, keeping the included angle consistent. Repeat this from the other end.

Now, depending on the hardness and texture of the stone, cut along this line again, barely touching at the point, increasing the depth to the centre, repeating until the slopes are filling the space, cleanly tooled and with a sharp intersection at the corner.

The curved line will, at first, be a bit fiddly until the knack of blending in the approaching slopes is mastered by rotating the head of the chisel whilst chasing; the blade will then cut progressively more at its outer edge whilst moving along the slope.

The other deadeye will have all sides curved, giving even more chance to practise.

Cut your mark on the joint and relax.

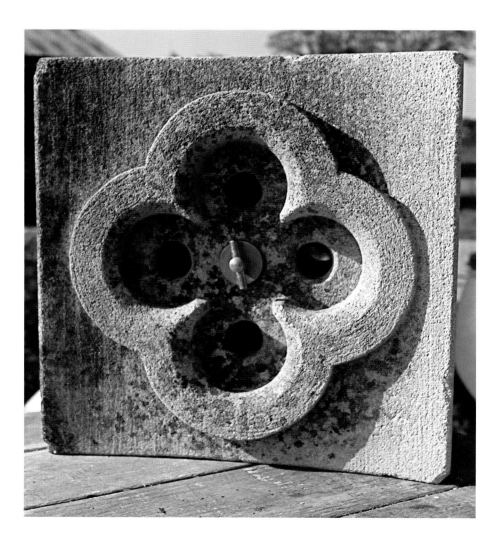

Ventilation stone in quatrefoil design to match building style.

CHAPTER 11

THE BALL FINIAL

Here is the opportunity to embark on one of the most ambitious, satisfying and seemingly mysterious of projects by creating a perfect sphere sat atop a piercing pyramid of gently curved sides – an unqualified product of intense discipline and amplification of tool skills.

ACCURACY IN PREPARATION

By now the importance of accuracy should be second nature, with the obvious realization that stonemasonry is not some practice of freehand sculpture; we are not releasing the 'hidden forms' from the within the stone. Rather stonemasonry is the transformation of geometry and design into three dimensions. To produce a sphere from stone requires that the switch from two to three dimensions is a dynamic process controlled by design and craft.

The block to cut this from – a rectangular block, defined as 'square' in masonry terms (to allow for the pyramid socle) of square plan – should be perfectly cut; otherwise there will be a lot of time wasted attempting to get the sphere perfect. Placed in a cube, our sphere will come into contact with all six sides exactly in their centres, and it is from these tangential nodes that the working starts.

SETTING OUT

The templet for this piece would be best fabricated from zinc, as this would allow it to be used without worry of distortion

when scribing around; also, as a single ball finial is unnatural you may need to balance it up by producing its twin, to confirm that it wasn't a lucky accident that produced the first!

Apply the templet to the four sides of the block, with the top of the ball level with the top of the stone and the outside

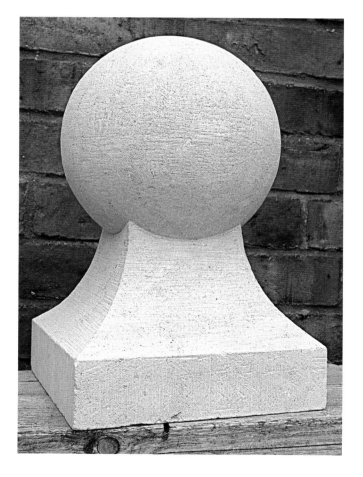

Ball finial.

OPPOSITE: The sphere – of all the shapes in geometry, it is probably the most perfect and is fittingly used in this role as the culmination of an architectural element.

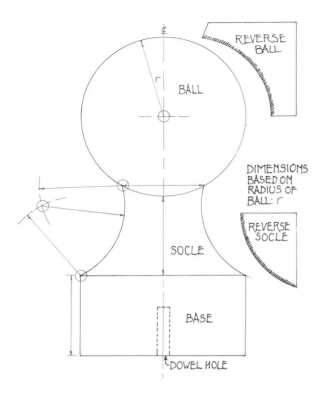

Drawing for templet.

Labels in drawing:
REVERSE BALL
BALL
r
DIMENSIONS BASED ON RADIUS OF BALL: r
REVERSE SOCLE
SOCLE
BASE
DOWEL HOLE

larger than the sphere it will contain, and it is then marked up with a vertical centre line on each face carried over the top. The templet is centred on this, with the top of the ball just below the top of the stone. This allows for discrepancies in the stone, so a small worked square is cut into each face on the (five) extremes of the sphere, and the centre is marked on these. These are the levels to work from; the advantage here is that cutting a curved surface to blend into this point is somewhat easier than blending onto the flat surface of the block, as the nodes can easily end up as noticeable flats with the exact block.

Alternative to Sphere Templet

If the block is true and square in the cut it is perfectly feasible to dispense with the bother of making a sphere templet (this one time – but it is not common practice) by scribing the

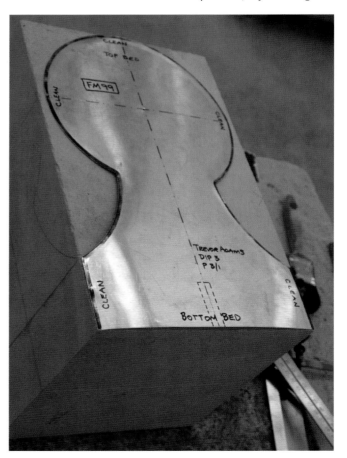

Templet applied to stone.

edges meeting, infinitesimally as tangents, at the corners, as do the socle mitres (for expediency the socle should have the same plan as the block).

With this scribed and pencilled in, the extremities of the sphere need to be determined. Scribe 45 degree diagonals from all the top corners of the stone across the marked spheres (on each side) and on the top; where these cross are the tangent points. To double check these are correct use a sinking square set to half the size of the sphere (one radius) and check that lines meet up all around the block, through the intersections.

Alternative Block

In the probable chance that you may not be able to get a block cut to the exact dimensions for a sphere, where precision is paramount, there is a slightly more practical approach that is taught to apprentices. The block is always cut slightly

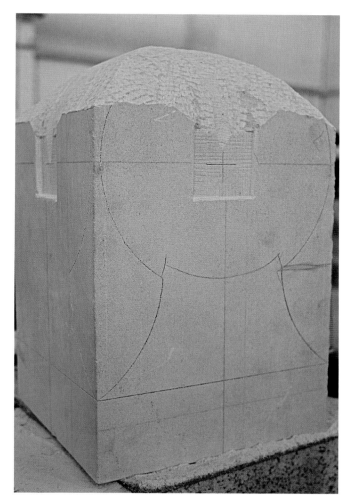

Levels cut in and top being roughed over. All the markings on this will be lost as work proceeds.

This is not a large piece, and it will often be advantageous to lay it on its side to work the curves. You will probably find a personal stance that allows for the most effective working.

Moving the stone about will rub out pencil marks, especially on the face resting on the carpeted banker, so make tiny centres with a scriber in the tangent points of the sunken patches that can be quickly and accurately redefined with a pencil.

TAKING THE CORNERS

Laying the stone on its side, proceed to cut a draft that follows the circumference line running around the mid point of the sphere. Do this in sections from tangent nodes with the imaginary line (you will have removed the pencilled line by now) in the centre of the draft.

To get into the stone without jamming up and to allow accurate checking of progress, valleys will need to be created by cutting away the bulk on either side. This could also include the use of chamfering to get down to the curve, but it is more beneficial in the long run to work down in curved drafts – you can use claws for the initial roughing out, but always do curving drafts working down to the line in one sweep at a time till you meet the two nodes and the reverse fits.

These curves are checked all the time by applying the reverse in a radial direction from the corners of the block and removing stone until the reverse fits perfectly around the

shape directly onto the stone using dividers, springing from the intersection of 45 degree lines dropped down from the corners and in the centre of the top bed (obviously a finial will not have a top bed!).

There will still be a need for the reverse and socle templets, as with all curved surfaces it is necessary to have a reverse. The arc for this job needs to be a complete quadrant as it has to touch on each tangent point whilst you are cutting the stone; do not have a shorter reverse, as this is the only way to be sure of accuracy.

While the block is still square, turn it upside down and in the centre of the base bed drill a hole deep enough and of appropriate diameter for the fixing dowel to locate in.

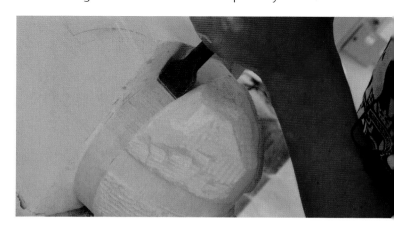

Working the circumference.

ABOVE: Taking corners off.

BELOW: Quadrant and circumference drafts cut with a bullnose.

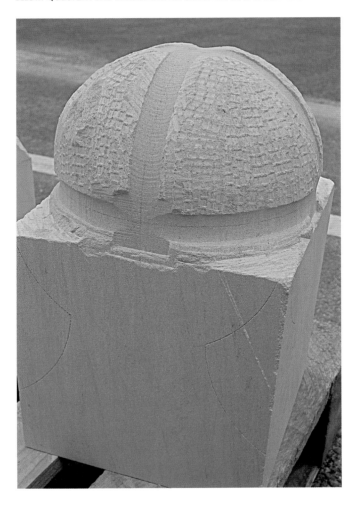

quadrant from point to point. The subsequent quadrants are cut in the same way, and the reverse should be able to slide smoothly around the completed draft.

Once the circumference is completed remark the tangential nodes with pencil crosses, then position the stone to allow you to cut a quadrant from these to the top of the sphere in the same manner.

ROUNDING OVER

As before cut a valley from the side tangential point to the top that allows access to the stone and work down to the line as before. The distance between these nodes is exactly the same, so when the extremes of the reverse touch the stone at the same time the curve will be complete (and touch all along its length).

Once the four drafts have been put in, the large clumps of stone on the corners need to be removed. If, at this stage, you want to remove all of this spare to get a shape nearer the finished surface, allowing more working room, go ahead; it can be handy to do but it will decrease the stability of the piece when laid on its side (this is a relatively small stone) – stability that can be useful while drafts are cut down to the intersection of the socle and sphere.

These drafts are cut by advancing from the existing, inching into the stone with the reverse resting on the top draft and gradually extending into the bottom half of the sphere. It is best to make a mark on the reverse, which indicates the extent of the curve before it intersects (hold it to the templet touching the socle, and mark where it crosses the equatorial line) and stopping slightly short of this.

Now punch and claw, to a surface well above the finished surface; then with a bullnose start to work drafts radiating in from the circumference line to the top of the sphere. The bullnose is to prevent any undercutting and will eventually give a hemi-spherical dome attractively furrowed from apex down to circumference; this is the surface to keep, right to the point prior to dressing the stone in. There is a practice of using a flat chisel to work this section that can be effective, but the use of a bullnose allows the control of following a line with the centre of the blade without going into areas to the side.

That was the easy bit! The next stage will be to carry the curved surface down from the equatorial line to intersect with the curved sloping surfaces of the socle.

A reverse for the curve of the socle slope should be made.

Ready for finishing top half.

MAKING A CLEAN START

Concave mouldings, such as these on the socle, are slightly different in their manner of working and can be spoilt visually by a 'soft' transition from flat to curve (that is, a rounded edge).

To give this its best start make a small pitch by placing a chisel or boaster into the scribe line and sharply striking into the curve, at right angles to the flat face. Restrain the travel of the chisel to prevent going in too far and damaging the curved surface; this is replicating the tangent strike in geometry. The result will be a crisp arris.

Cutting in the socle.

Checking with reverse.

Meeting the socle.

MITRES

From the section you will be able to determine where the curve intersects the sphere at a low point, but it does not indicate where the mitres will strike, as these are still within the stone. With a straight edge as guide join the two sides up by cutting the lower part of the curve to the intersection using a combination of straight chisel and bullnose; you can then continue the curve up using the reverse as a guide. Whilst this is happening curved drafts for the bottom half of the ball should be carried down to the notional bottom pole of the sphere and meeting the slope, resulting in the correct intersection of slope and sphere – this will develop by working down till precise cutting from one curve to the other creates a corner that slides up the socle slope; it cannot be marked out beforehand.

To cut the other slopes that join up the two sides already worked, a mitre must be marked on the slope by using the mitre board set from the apex point to the corner of the socle, which is carried out as for a standard mitre but with the added excitement of not having a flat surface on the top. So con-centrate and double check – measure twice, thrice and then cut once.

A concerted approach to cutting the slope and the sphere intersection, diligently using the reverses and straightedge, will result in the final shape coming into existence.

FINISHING

Whilst the ball is now fully formed the surface will be noticeably broken into different areas of working with the bullnose cuts showing the direction of that session's cuts. So now it is

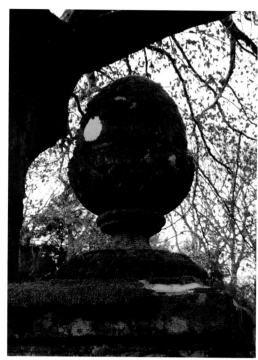

necessary to blend the surface into one finish. The practical way to do this would seem to be to rub it down with abrasive paper, but as explained previously this will give an unhealthily smooth pallor to the stone – we want a surface that invites the eye in to stay, not to slide off. What is now needed here is to take a half inch chisel and, working down from the apex regularly in all directions, take the final surface down with fine, regular concentric tooling right down into the intersection – something you cannot do effectively with sandpaper.

Touch

There is an extremely effective method of checking that the flow of the curved surface is flawless, and that is to simply run your hand over the ball; dips or bumps will be felt quite easily – though you will probably have been doing this from the first bullnose workings – hard not to!

Pragmatically while aiming for perfection there may be surface deviation, but if imperfections can only be felt but not seen then they should not be an issue this time. The tooled finish will help minor imperfections blend in, whereas a polished finish will expose them unless it is absolutely perfect – and then the unnaturalness of the finish will mar the final image.

Socle Tooling

What will be noticeable is that whilst you have created attractively tooled surfaces, the straight faces of the socle will probably still be smooth from the saw cut, or have saw lash. Though these will contrast strongly it is a personal (but wise) decision to drift a boasted surface on perpendicular to the base bed; optically this will make the base look narrower and taller, giving the ball more emphasis.

MASONRY SKILLS ACHIEVED

You have reached the point on this worthwhile journey that will allow you to understand and undertake most of the masonry work that is required in construction. More complicated pieces may also be achievable, as it tends to be the geometry and setting out that is the difficult part, which once marked on the stone only requires the stonemason's practice of cutting to the line.

CHAPTER 12

SITE PRACTICE

You will have noticed by now that working on site fixing the finished piece of stone into an old building is not always going to be a safe, comfortable and productive pastime. We will assume that these tasks are going to take place from well designed scaffold suitably placed for high level insertion, the ladders are all tied up and there is a weatherproof cover over the work area. You need to consider the possibility of specialist equipment and the aspects of good working practice on site.

EQUIPMENT

Set-up Costs

It may seem to be throwing money away to purchase good quality machines when you can work and move materials by hand, or to procure a top-rate sheeted-in scaffold to get to an area accessible by ladder, with expensive sheeting to protect against the elements that are probably never going to prevent you from working and anyway there is always a good water-proof jacket, so why bother?

Firstly weigh up the costs of providing good access, adequate shelter and useful plant against the amount of (money earning) time lost working slowly from an awkward position, battered and frozen by inclement weather or doing by hand tasks more suited to a machine. Only you can be the judge of whether the rewards are worth the investment.

OPPOSITE: The exposed Hardy's Monument, overlooking the Jurassic Coastline, is typical of the location the stonemason could be working in. Far from the madding crowd, it is essential to be in control of the project, and to assume all the responsibilities and requirements necessary for a safe and productive working environment.

New finials, Chittlehampton, Exmoor. These finials replaced inferior stonework that, while only ten years old, had blown off in gales; they had been built and fixed badly using sub-standard building practice and inappropriate techniques. The earlier ones, fixed in traditional method more than 100 years ago, had withstood the weather.

Sheeted-in scaffold. With a roof this allows work to carry on whatever the weather.

Lifting

Can the stone be delivered to the area without damage to it, yourself or the building?

Climbing ladders with any stone larger than a bowling ball is not feasible and also dangerous, so get hoisting apparatus set up appropriate to the weight of the stone, handily positioned to lift from an area easily accessible for transporting stone to a position on the lift that gives unhindered access to the workplace. Remember all lifting operations will need at least two people to do the task – especially when using machines.

While it is perfectly acceptable to lift and move large stones around by manpower or rollers, to get these stones into position is going to require something more.

Straps

We are assuming that there is a powered crane or hydraulic lifting arm available, large enough to manoeuvre stone from transport to scaffold, and also for setting into position.

Stone will usually travel or be stored on blocks that allow straps to be slung underneath and attached to the crane. These are useful up to the point that a stone needs to be set into place, where it would be difficult to get the straps from below the stone. Placing wedges is fine as long as they do not damage the arrises, dislodge the stone when being removed or get fingers trapped. An old stonemason's trick was to place

Hoisting stone: one man at the bottom controls the rope to prevent damage.

sugar cubes in the joint, allowing the straps to be pulled free and with a squirt of water the cube dissolves and the stone settles into place!

Lewises

Good quality straps were not widely available in far-off times, and even today they have limitations; the answer is to use the lewis. A lewis comes either as a split-pin lewis or as a three-pin

Three-pin lewis. Slightly complicated, but used correctly this is the safest way to hoist stone.

point into a stone. The stone has a dovetailed slot put into the top and the lewis is assembled in this before lifting. To remove it the assemblage is dismantled in reverse order.

Working Space

Have an area to store the stone, while it is not needed, that is located out of the way (on ground), level and can be covered.

If you have a portable banker it would be prudent to set it up nearby for fettling the stone. Alternatively, if possible, this banker could best be placed next to the insertion area. It is always sensible to reduce moving or double handling the stone up and down, and prevent a situation where you have to foolishly try and work the stone while it is resting on the scaffold planked floor. So when planning the scaffold make sure there is enough room on the lift for all this activity, and more.

Mortar and Sand

Unless using ready mixed mortar, there will be dry materials on site that need to be protected from water (until ready to be mixed), so put a pallet down to stack the bags off the ground and with waterproof sheeting create a wrap that can be secured over the materials not in use.

Putting all dry materials in storage bins with lids is a better proposition, especially if there are excess materials that need to be moved around. They can be stacked outside without taking up workshop space, whereas bags are awkward to decant materials from, will split, leak and generally create mess.

Designate a mixing area for mortars where, if using a powered mixer there is a stable surface, with any cables safely secured out of the way.

Ensure that sufficient water is available – and more importantly that there is a waste disposal proviso for contaminated water and all other rubbish from the site. There is always a prodigious amount of mess from even the smallest preparation of mortar, so it is best if you can efficiently clean it up afterwards. Cover the ground with board laid on polythene sheeting and always confine the work to this area.

Buckets, sponges, brushes and a shovel should always be at hand. Remember to clean them thoroughly every time after use, as the build-up of old mortar will contaminate new batches and quickly reduce the useful life of tools.

lewis. Both are fitted into the top of the stone and allow the stone to be suspended without any accoutrements hampering the siting of the stone directly onto its bed.

The split-pin lewis goes directly into a drilled hole, and when pulled it will lodge firmly (the heavier the stone the firmer the grip) allowing the stone to be lifted easily, spun round and manoeuvred with no exertion. Slackening the tension allows the pins to be removed simply and quickly.

The three-pin lewis, sometimes known as St Peters Fingers, is probably the most secure method of anchoring a lifting

Banker on scaffold. The scaffold platform was built to accommodate working areas, as the amount of dressing-in here was extensive.

Sand (aggregate) should also be covered up, to protect it from bulking, to keep it from contamination from site rubbish such as stone fragments and to lessen its appeal as a toilet for local animal life.

Personal and Personnel Safety

Safety considerations require that there should be equipment on site for minor injuries and emergencies; this should include eyewash, plasters and a decent first aid kit.

Also be aware that working on your own with heavy materials and power tools is not recommended, without the ability to contact help, or someone knowing where you are and what your movements are, in case you do not come home – stonemasonry is not mountaineering, but in principle the risks are similar, with the added hazards of electricity or caustic materials (and the irony of being near to civilization).

WORKING PRACTICES

Pulling It Out

The scene is set: eager with tools in hand, your freshly masoned piece of stone is ready to take the place of the dying fragment of another age, bringing the façade or whatever back to life. At the moment it is sat on a bench next to the condemned stone. What now?

The old stone must come out with the least disruption to the surrounding stones; these should not be involved or you could end up with a major rebuild job as dislodged blocks tumble into the fray. Preventing this can be accomplished by (carefully) destroying the stone in situ or by lifting it out as one piece.

Lifting out is best as there is very little banging or smashing necessary and little need for power tools (the use of power tools can set up vibration that may loosen surrounding stone as mortar is shaken out of the joints). Luckily if the mortar is

Stone ready in the yard at Salisbury Cathedral, for a repair phase. These are all the same, so accuracy is essential when working.

holding the stone in place even though, frustratingly, all the mortar has been removed; use a fresh unworn hacksaw blade to cut through this, not an old one that is blunt from chewing through stony mortar or you will be there forever. Powered tools that look like they can do this do not tend to have the thinness or length of blade to get into tight joints, and hitting obstructions can bend blades or jam up, so be wary.

If force is necessary to get the stone out then chiselling with care is the way, with some help from power tools. With a disc cutter it is best to cut out the centre of the stone by making relieving channels to the edges, stopping at the joints, and then collapse the remains into the centre with gentle pitching, trying to keep the walloping with huge tools to a minimum.

Drilling holes across the centre gives the removal a different aspect as the depth can be reached and it is more controlled; depending on the stone you will decide on the manner of removal and best tool.

soft enough for this to happen it can be removed by using a saw blade from your toolkit collection to break up and drag out the mortar in the joints and beds (remembering to protect your hand by making a gaffer tape handle). This subsequently allows the stone to be pulled out – do not lever it out using surrounding stone arrises as fulcrums.

It sometimes happens that there is metal dowel or cramp

Checking for Fit

After all loose and unwanted material has been cleaned out of the wall-space, the replacement stone should be offered up

Cutting out stone with disc cutter. The badly eroded stone is crisscrossed with cuts, and these can be knocked out with less disruption to the surrounding masonry.

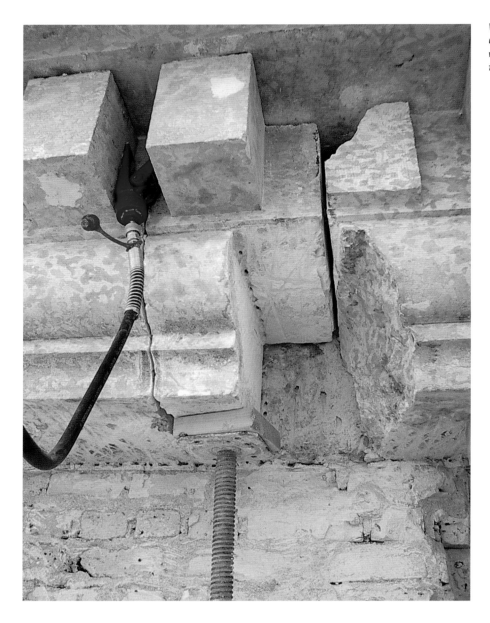

With joint cut, stone is jacked out of position using a hydraulic wedge; note the support, with protection for stone, to prevent it falling out of control.

for checking the fit by sliding it into place; before attempting to place it, double check all measurements so that it will not snag and break off delicate edges.

If the stone is to have a nice tightly jointed fit, which is always preferred, trying it for size will mean difficulties in getting it out again without damage. This can be overcome by using a strip of plastic templet film or similar looped around the back of the stone enabling it to be drawn out easily (do not use normal paper as it will not cope with the weight, and wet stone will make it tear easily. The band material used to tie stone down on pallets is also good for this, and sliding the stone in on top of straightedges or a sheet of zinc are also workable solutions.

When dry fixing or checking joint spacing, small softwood wedges can be temporarily placed jutting into the bottom bed whilst manoeuvring the stone around. It can be decided at this point whether spacers to support the stone at the right height are needed; these can be plastic packers, wedges of slate shards or best of all lead packers made from a strip of lead sheet, which if necessary can be folded over and then adjusted by flattening with a hammer to the desired thickness.

DRILL BITS

The quality of drill bits available varies widely. It is best to buy recognized names rather than cheap ones, as they do last longer and will do the job.

Masonry drill bits have a tungsten tip and rely on grinding away the material, which is really only effective when used with a hammer (percussive) action; this can be detrimental to delicate structures or small stones. Masonry bits are also not good for drilling completely through stones, as they will blow a spall out the back.

On the harder stones masonry bits are the only choice as the other type of bit, high speed steel (HSS), loses its cutting edge rapidly as it heats up. HSS bits are primarily designed for metal and softer materials, and drill by cutting through the material. They can be used for medium to soft stones but need to have their edges sharpened frequently.

With experimentation it is possible to find the best bit for the application; it may also help to alter the angle of the cutting edge when sharpening on the grinding wheel.

ABOVE: *Hank offering stone up for size. (Photo: Ian Constantinides Ltd)*

BELOW: *Dutchman ready to go into a cornice.*

Stones going in; note they are dampened to prevent suction of the moisture from the mortar.

New stone with a tooled finish to give character. Neat joints and clean work are both necessities.

Wedged in, checking for fit and lining up. Here is where guesstimating comes into its own as the line is lost in the original, through settlement and past repair, so the replacement must fit to the best of on-site judgement.

Fettling in situ. Here Ian is removing stone at the back to make the new piece fit.

TO THE FUTURE

By now, you will probably have an interest in past work and those who carried it out, but the information to give these a real presence is scant. Here is a proposal to halt that trend and make yourself known to future generations.

Before the stone is fixed, get a piece of lead sheet about A5 and clean it off thoroughly with wire wool to a polished surface. Now, using a sharp scriber and clear lettering, write down your personal details, information on your family and colleagues, the owner/patron of the building, the type of work carried out and, of course, the date on the lead. Give it a coat of oil, fold it in half and gently flatten it, then put your mason's mark on it and the words 'Read Inside'.

This placed in the void at the back of the stone will, at some time in the future (20, 200 or 2000 years??) give historians something to muse over, allow them to date and recognize your work and perhaps let your descendants know of their worthwhile ancestor.

As an archival document it cannot be bettered.

Putting It In

The stone should, all being well, be a perfect fit but this is not always the case! So any necessary dressing back should, as much as possible, be carried out on the banker to prevent the stone being jarred after fixing. (Obviously it may need material removed from the back if it sits proud or the space it fits into appears to have got smaller since the measuring stage; I always measure for an exact fit, knowing it is easier to remove stone than add it.) If the stone is soft a coarse file or drag can remove material. Harder stones can be dressed off with a scutching hammer, chisels and diamond pads to remove high spots. By rounding the sharp arrises to be hidden in the wall, you will enable mortar to slide around the stone.

The stone is now ready to be fixed.

Fixing

Dampen up all surfaces, by mist spraying to saturation, well in advance of fixing to prevent the dry stone sucking out the moisture from the mortar. Too much suction results in lack of

CURING IN A NUTSHELL

Using the correct mortars for stone construction will mean there is no cement present at all. Thus the initial set will have to be controlled, as lime needs to slowly lose the moisture within, which will allow the growth of the binding crystalline structure to take place at the right pace (this is true of cement as well but ignored by most – hence premature shrinkage and cracking of this hard material).

Lime that dries too quickly, evinced by it becoming powdery and noticeably whiter than correctly cured mortar, should be considered to have failed; if this happens it should be removed and replaced.

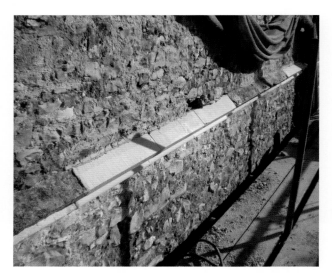

ABOVE: On street repairs, Aix en Provence. The idea here is to rub back the stone, once fixed, hence the messy mortar on the stone face.

LEFT: Fixed stringcourse. The tooling on the upper surface will actually help shed water.

manoeuvrability, possible jamming and subsequent failure of the mortar if it dries out too quickly.

Also ensure that the face of the surrounding stonework is damp so that limestains can be lifted off efficiently, as once dry they are near impossible to get off. Use covers against the wall so any mortar dropping out will not land on stone below.

The stone to be fitted should be thoroughly damp but not running with water; some porous stones need to be immersed in a tub of water then left to dry out slightly – but not too much!

JOINTS

The width of joints is going to be an important consideration when placing the stone. We can assume that large joints are easily packed out from the face after the stone is placed, so this note is not for them.

With the fine jointed stone, on the other hand, mortar needs to go in at the same time (as the aggregate will be finer it will flow better especially if the ratio of lime is high) so place enough mortar into the back of the hole to fill that area and butter up the sides with smooth quite wet, but not slump-

THE PROBLEM WITH LIME

Splashes from the mixer will contain lime that can irrecoverably stain stonework, so ensure that the mouth of the mixer is facing away from the building. If there is a chance that splashes can occur have a hose/sprayer to wash off immediately.

Never allow the lime-laden water to stay on the stone, or even worse run down the face of the building, because it will look unmarked while the surface is wet, but when it dries it will be bleached white with insoluble, impossible to get rid of stains.

Always wear eye protection when mixing mortars, as lime is a caustic material. There will always be dobs flying around, and if one lands in an eye it is an immediate trip to the doctor or hospital.

Gatehouse, Corfe Castle. Using inappropriate materials and building techniques in restoration work has caused calcite build-up from leaching lime (white staining on the soffit), where the mortar used was not able to cure properly and subsequent rain has washed out the lime. The fake (exposed) rubble interior here is not appropriate; if they had only looked, they would have seen that the original rubble infill was built up in rough courses concurrent as the dressed stones were fixed. Note the differing styles and lack of adherence to one style – indicating lack of discipline and understanding of masonry structures.

Castle Drogo, Devon. Lime leaches out of the joints here, allowing water ingress. This is granite masonry, and the use of slow setting mortars has caused many problems.

Dutchmen positioned with mortar, consisting of lime putty, squeezed out and drying up nicely, ready to cut off neatly.

ABOVE: Chopping out old stone ready for a new piece. The joints around are raked out ready as well.

BELOW: Pieced into the cornice, the new work has a textured finish and starts to blend in quickly.

A cornice piece inserted with dry joints that will be packed out later.

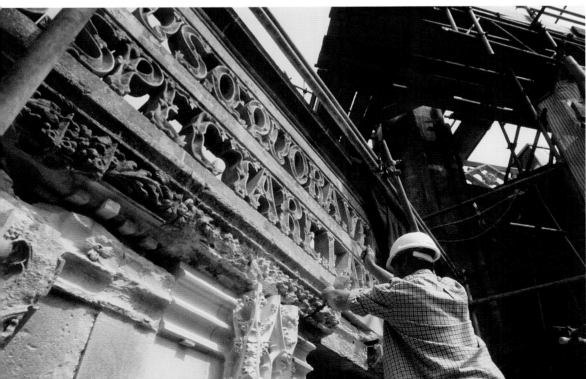

Grouting in stone. A wire is poked down the tube to agitate the mortar and help it flow. (Photo: Ian Constantinides Ltd)

ing, mortar and without wasting time push the stone into place. Resistance as the mortar thickens up will be overcome by placing wood on the face and tapping it with a mallet or hammer till correctly placed (the wood will spread and soften the force and allow quite fine adjustment).

There should be sufficient quantity of wet mortar to have oozed out onto the face as the stone is pushed in; if the consistency is good (piped icing or toothpaste) it will hold itself off the surface. Neatly, without smearing onto the stone, cut off the excess by running along the joint with a small trowel and lift off any lumps. Then clean any stains off the surface by swabbing with a partially wrung-out sponge washed continually in clean water.

Filling Joints

Pack any open joints with mortar (keeping it off the face of the stone) and if necessary place wedges in to keep the stone in position whilst curing. Do not use wood; rather use slate or lead set as deep in the joint as practicable.

It may be that the back of the stone has not got enough mortar in, due to difficulty in packing out; or because of the size, it was set in dry then pointed up so the recesses will have voids (no filling). This is overcome by grouting. Start by drilling through a joint above the stone, preferably into a pre-made joggle. Then a tube with a funnel attached is inserted for grout to be poured through to fill out the voids. Be careful not to overload the new joints with so much pressure from the grout that they get pushed out; best practice is to grout when the mortar has taken at least an initial set.

Cover the fixed stone with hessian under plastic sheet and leave it to cure for as long as necessary (more than five days) before carrying out any dressing in situ.

When dressing in the stone always use minimum force, sharp chisels and strike towards the centre of the stone if it needs chiselling, if you are dragging the surface back, work inwards from the joints to prevent spalling of the arris.

Added Steel

In some situations it may be necessary in addition to mortar and packers to fix the stone using dowels or cramps to prevent it falling out or tipping over, or if it needs to be part of a reinforced system such as handrails or copings.

The recesses for the cramps need to be cut in with minimum disruption to the fixed stone, so if possible place the

Setting restraints and dowels. (Photo: Ian Constantinides Ltd.)

stone in dry and mark up before taking it to the banker to cut out, allowing enough depth for the cramp to be covered with mortar if necessary.

When the set is cut it should be flushed out to remove any dust, as this will prevent the mortar attaching to the stone. The hole is partly filled with a mobile not-too sloppy mastic mix; the cramp can then be gently tapped into place, and any material squeezing onto the dampened stone cleaned up quickly. Mask around the cramp with tape and sheet to prevent the mastic getting onto the surface of the stone.

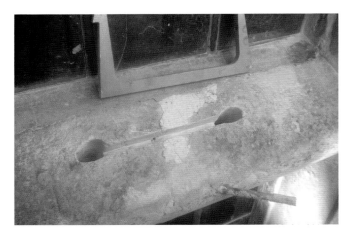

Holes and cut-out prepared by cleaning out of dust, and now the cramp is ready to be fixed.

Rusting iron cramp across ashlar blocks has expanded, which causes enough pressure to split stone. This shape fracture is typical of this problem and a good indicator.

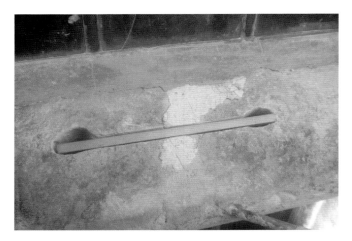

Cramp dropped in to try for size; this is to strengthen a transom stone that has fractured.

A stone from Salisbury Cathedral spire, showing cramp and keying cuts to lock in place. The spire is a delicate structure; rising to 404ft (123m), it is a cone of thin blocks (some only six inches thick) all held together with the shown techniques. Interestingly by modern structural engineering standards the spire should not be capable of surviving (just as a bumble bee is theoretically incapable of flight), so a hi-tech armature, costing millions of pounds, was inserted barely 700 years after it was built.

Dowels

Dowels, set into the wall, are used to support stones that overhang or have a lot of mass outside of the wall line. The material should, as in all these uses of metal in structure, be stainless steel of a marine grade quality (also classed as food grade), with threaded bar or reinforcing rod, as they both provide a means to grip the mastic setting them in; smooth bar should not be used.

Obviously holes for dowels should mate up the stone to the wall accurately. A simple method is to cut a cardboard template the size of the back of the stone and by applying this to the stone mark the dowel positions by poking a hole through; then apply it to the place it needs to be fixed into and by marking through the holes with a pencil, it is easy to see where to drill.

Cut the dowel(s) shorter than the combined length of the hole and fix them into the wall using the preferred mastic – preferably a lime-based material or a modern resin that is not affected by moisture. Remember that the stone must be slid onto these dowels, so make sure they point in the direction

Dowels for locating and tying back mullion face replacement. This type of repair, commonly used on stone windows, is known as halving, where the old stone is cut back to the glazing line and a new piece fixed on the face. It is effective for windows that erode on the exterior as they can then be repaired without dismantling the whole piece.

of insertion and always try the stone in dry to make sure the dowels will locate correctly before the mastic sets. It is good practice to have the stone in place while the mastic cures; otherwise you may find the dowels slipping out of alignment, entailing more drilling of the stone. (The two-part resin marketed as stone 'glue' –often called Akemi – and widely used in stonemasonry, is usually a polyester resin that does not set properly when damp, as the stone on site will need to be, so try not use it.)

Secret Fix

A method that will lock a stone solidly into position after it has been placed, is to drill a hole in each vertical joint from the face to the back inserting the dowel set in mastic into this, and then pointed up; the hole need not be overly obtrusive as this only needs to put a solid lock in either side and can be effective when only slightly wider than the joint itself.

Carry on fixing the stone in the usual manner, with the

holes for the dowels filled with the mastic; be aware of all the necessary setting time limits of the materials used here and coordinate accordingly.

Or Traditional Materials…

If the use of such a modern material as stainless steel does not sit comfortably with your approach to traditional repair then there are alternatives:

Slate can perform practically as dowels and cramps if it is cut correctly with the cleavage plane running along its length to provide shear resistance. Whilst the size will be substantially greater than the metal alternative, the effect can be considered most fitting and also cause less trouble when maintenance work is carried out in the future.

Small pebbles can be slipped into holes drilled in joints as a lock similar to the one mentioned before.

Other non ferrous metals are usable – though be aware of their possible reaction with the moisture in the stone as this could cause staining such as verdigris (copper-sulphate).

Wood should only be used if it is considered tempo-

rary until mortars set and allowance is made for the wood rotting away or swelling when damp. In dry climatic conditions wood can be used; the Acropolis in Greece had oak plugs that were used to locate drums of the columns and have survived through the millennia to be discovered recently, only feasible if they can be isolated from moisture and the atmosphere.

Do not use terracotta or stone for dowels where there is any chance of load in a direction that could cause a shear break.

When iron, usually referred to as ferrous fixing, was traditionally used in masonry construction it was known that it could only be viable if isolated from the harmful effects of moisture, usually by setting it completely in lead, either while the lead was molten or by grouting around the iron. This works well if done correctly but can cause serious damage if it is not; a lot of repair work to historic structural masonry is to stone that is affected by the massive expansion of rusting iron which ruptures and dislodges stone.

Dowel material (from top) Stainless steel threaded bar, stainless steel re-bar, stainless steel bar, brass bar, copper bar. Note the staining on the lower ones; this is the reaction with air pollution/moisture, and the result can be verdigris (copper sulphate) staining to the stone. Interestingly traditional biocides used copper sulphate to prevent and kill plant growth, and so an early indicator of verdigris staining can be a lack of algal growth below the area where the material is washing out.

The age-old problem with rusting metal.

GOOD HOUSEKEEPING

Keeping It Clean

When removing or working stone on site, try to have a catch sheet for the rubble and dust, especially if on a scaffold lift above others, and clean away mess regularly.

If there is a chance of rain or running water through the work area, ensure there is no uncured mortar in the mess, as this will stain below when it is washed out.

When leaving work covered ensure that the sheeting is securely tied down and that water cannot get in to wash out joints, because if it does it means mess.

Never pour liquid down drains or keep materials in unmarked containers.

Dispose of all waste correctly and tidy up after the job is finished.

Keeping it Safe

Never throw anything off a scaffold.

Do not leave power tools or leads out where they can get wet, or let cables be damaged or become covered in debris. Electricity rules on site are different from what is used at home – they are for your own safety so follow them.

SITE PRACTICE CHECKLIST

Working on site puts the artisan on show, and at the same time makes you responsible for yourself as well as others.

So always:

Plan (the task)

Price (to get the job and earn enough)

Purchase (enough materials)

Prepare (the site)

Proceed (with the work)

Polish up (after).

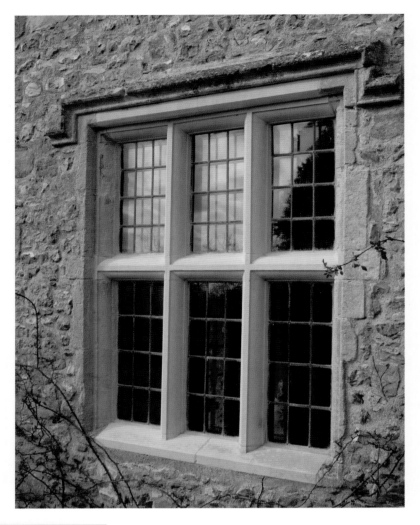

Replacement mullions and transoms in Beer stone, by the author.

New stone should always be to the original design and not be artificially 'blended' in. If artificial weathering is applied, at some point in the future there will be no physical record of what was originally there.

Dress Sense

As well as wearing all the appropriate protective kit, you will do well to consider what constitutes general work clothes. The work here on site (and in the workshop) is dusty, and when water is factored in, quite filthy. It would be unwise to wear the clothes you work in as you climb into your car, sit on a bus or relax in front of the television, as there will be unnecessary mess.

Get some overalls and only wear them while working; this means as time goes on that you only need to replace them and not whole outfits; you will also not destroy washing machines on a regular basis and you can use the car at the weekend – your choice.

If tradition is a strong influence, why not get an apron – it doesn't have to be leather. Aprons are handy, easy to remove and provide a fair amount of protection to clothes.

WHEN IT'S YOUR FAULT

The only real problem that stonemasons have is when the stone breaks – usually after considerable work has been put in. Guidance to some basic repair techniques is given in Chapter 14.

This is not the end...

CHAPTER 13

ASHLAR BUILDING

Take a walk along a modern street, wander up to a building and look at the construction quality of the walls. If not covered with render the brick or concrete block will be displayed roughly set in ugly looking mortar, with large, uneven joints that allow the arrises to wander up, down, in and out, and possibly finished off by some hideous staining. All in all this is a sad indicator of how we neglect quality, for in general with modern masonry the need for economy, speed and the lack of skills by the designers and the builders usually results in a travesty of what a masonry wall should be.

In this chapter on the techniques of ashlar construction we will learn how to produce work that will compare well with high quality traditional work, work that was produced by skilled artisans when awareness of quality and standards required an ashlar wall to be almost a work of art.

RIGHT: *Precise and technically perfect Victorian masonry. Fine joints and good design show the beauty of good stonemasonry and fixing.*

OPPOSITE: *Here a rubble-stone wall has been bought to a neat termination with the fixing of ashlar blocks.*

LEFT: Rusticated stonework surrounding a niche. Here, as with most unskilled repairs, the unsightly mortar is too hard and of the wrong texture to be correct for this type of work.

ABOVE: Masonry in Winnipeg. The masonry here has been covered in a cement render to 'repair' the eroding stone, which has actually exacerbated the problem with hideous results.

Church, Lalibela, Ethiopia. Incredibly this and other examples, such as Petra in Jordan, were carved from the solid rock – imagine the setting out and templets! (Photo: Niall Finneran)

MONOLITHIC CONSTRUCTION

When a grand building was commissioned in the past the idea was to produce a well-designed display of wealth and importance, which meant using rare or expensive materials (or at least appearing to) and that has always meant stone.

Stone has continuously been available but often in small sizes or of poor quality. While this was fine for rubble walling of lowly cottages, or hidden around the back of grand buildings, when someone wanted to show off, walls of huge, if not single, blocks were desirable– but not always feasible.

Competent masonry design relies on the stones forming a solid unit, with the mortar as a gap filler providing pressure relief between brittle stone edges and allowing the inevitable settlement loads to be dispersed in the structure. This is rather different than modern techniques that rely on the brittle mortar sticking the blocks together, usually fracturing (basically failing) when any settlement occurs (and it always does).

So to get fine joints relies on regular blocks, quality fixing and more care in working; when this is done correctly the wall becomes one piece and presents the desired monolithic façade.

Curiously, when rubble stone was the only material available (affordable) then it would often be covered with a stone-coloured render lined out to imitate stonework representing solid block.

RIGHT: French country church showing typical treatment of masonry. Dressed masonry is used for the important parts and details, while walls constructed from local rubble stone are rendered over. This is how the majority of rubble masonry would be finished, so consider once again the incorrect appearance of so many of our vernacular buildings, with quaintly mottled rustic stonework. The original construction would have been at the least limewashed, and at most, rendered and lined out to represent ashlar.

BOTTOM LEFT: The sea wall at Teignmouth. The Victorians built miles of this type of structure, using massive barely-worked blocks of whatever stone was available.

BOTTOM RIGHT: The New Court, British Museum. With fine joints and clean ashlar surfaces, this gives an impression of huge stones.

ing with 1–2mm joints any discrepancy in height or with faces out of true will cause disharmony or, at the worst, a lump in the wall.

GOOD FOUNDATIONS

The base is the most important part, and it is assumed that, as we are building onto solid foundation (or an existing wall), it may not be perfectly flat. Therefore this must be brought up to level.

Stone walls in building should always have horizontal joints and never follow the contours of the ground – up and down bedding is a style that can only be used for certain drystone walling.

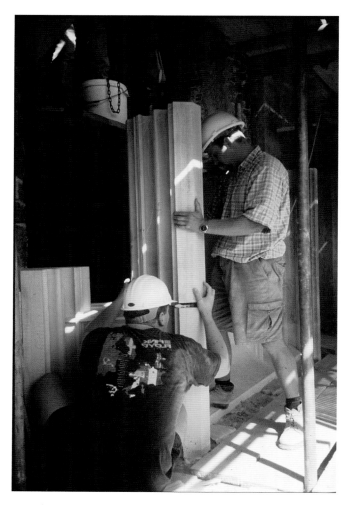

Many hands make light work. (Photo: Ian Constantinides Ltd)

String-line. Pink string shows up well against stone.

SETTING UP

You may find it useful to have a partner when using stone for ashlar, or any building, depending on the size of the stones used; moving blocks into position and getting them right will usually require more than two hands. If you can lift the stone you will need one partner; if it takes two to move the stone then three people are almost essential; bigger still will require lifting equipment.

All the usual fixing kit (described in Chapter 12) is required, as will fine mortar (usually an equal mix of fine stone dust and sieved lime-putty) and obviously squared ashlar blocks with joggles and fixing slots cut, of exact dimensions. When build-

The base mortar, if it is to level up a bottom joint, will probably be thicker than the ashlar, so it should be of coarse(r) mortar, to prevent shrinkage, with slate packers to make a level bed in a way that the bottom bed of the first course of stone is taken from a level of the highest point of the foundation course. It may be that regulations or practicalities necessitate a DPC be inserted at this point, which will mean a thicker bed still.

Using battens or pins in joints, set a very taut string line to run exactly along the top arris of the first course (making sure it is secure as it will be lifted and pulled about during work); this is the point of reference you will work to, so make sure it is perfect.

JOGGLES AND CRAMPS

A perfectly fixed stone will need to have all its joints completely filled with mortar to lock it into place. The nature of lime mortars means it is almost impossible to place narrow jointed stones if there is mortar on the vertical joint, so a V section channel is cut down the joint surface of the stone with an entrance hole at the top bed. Thin grout is now poured in this to fill and spread through the joints, using a long thin blade (hacksaw is good) in the joint to agitate the mix while it goes in, ensuring it will spread uniformly throughout the void. If the pressure is too much the grout can burst through the mortar causing a horrible mess, so consider this and possibly let the mortar gain a set before grouting. Stones that require to be locked into place with absolute certainty would have molten lead poured into the joggle.

Cramps are used extensively in stonemasonry as restraints and ties. The cut-out for these should be made before the stones are fixed, as working the blocks in situ could dislodge them (though I have designed a stainless steel cramp that can be inserted

Lead-filled joggle in a stone removed from the building.

into a cut-out made in situ with a drill and small disc cutter that requires no hammering of the stone). They should drop in easily and are grouted up with resin or mortar. Traditionally cramps and certain joggles of stones were set with molten lead.

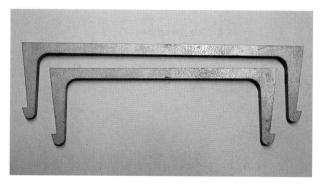

Specially designed stainless steel cramps, designed by the author, for regularity and strength. These are easy to prepare for and can be provided in varying sizes as required. Usually a fixer will bend a rod or bar on site, or have bent bar provided – producing irregular pieces that can vary wildly in style and quality.

Lead encasing a cramp. The cramp is stainless steel, but in this particular exposure mortar was not going to be appropriate so I encased it in lead.

BUILDING

Place some thin softening on the bed and place the first stone into place for a dry run; set it level and upright with fine wedges. Work out how much mortar will be needed to set it level and carefully lift off.

Now, ensuring all surfaces are dampened – remember stones should not be too dry while being fixed – trowel mortar onto the bed with a sideways chopping motion to create a rippled surface that will spread to the right thickness when you lift the stone back into place. Using a lump hammer (pitching hammer is best) beat against a flat piece of wood securely held against the stone to achieve perfect position before the mortar dries and grips, checking all the time with a level that the face, joint and outer arris of the top bed runs true to the string-line. Some use a large rubber mallet to knock the stone about. Though it is a useful tool it can damage softer stones or spall arrises, whereas a clean block of wood held against the surface will not bounce, spreads the load and can be adjusted for stones with detailing.

Holding the stone in position (use wedging if appropriate) ensure the bottom joint is well packed and that there are no stains on the stone – as the mortar is squeezed out of joints, always cut it off by running the edge of the trowel along the joint (do not wipe it); then if necessary sponge off.

Offer up the next stone without mortar to gauge the bottom joint thickness needed to bring it level with the first. When satisfied, set it on mortar as before, ensuring it is as close as possible to the vertical joint thickness by placing it without sliding, as this will ruck up the bed mortar, jamming it into the perp. You will need to keep these vertical joints clear of mortar, as there may be some more jiggling needed until the first course is laid perfectly.

Once the full course is laid, point up the perps with a stiff mortar to a depth of at least 25mm, keeping the face scrupulously clean; the back of an outer skin can be done with less attention to cleaning, but still to an effective depth.

NEXT COURSE

Check the top of this course for any deviation with a long straightedge and level; if there are any high spots you will need to get rid of them. First make sure the problem is not easily rectified by reseating a stone; then proceed to using drags, fine boasting, abrasive or possibly a disc-cutter with grinding head to get the required level – whatever happens

behind the face can be worked around, but the arris of the top bed of the stones must be an absolutely perfect level straight line, so concentrate on getting this right.

Reset the stringline to the top of the next course.

Now clean off any mess from the top of the first course and lightly mark with pencil where the next complete stone will go, equally spanning the perp below; from this you can measure back to find the length of the (half) stone, allowing for joints, that will start off this course. Cut it to size and do a dry run.

When setting ashlar bed joints it is handy to set non-compressible packers of the exact thickness under the four corners of each stone (these are obtainable in the construction trade at whatever thickness required), setting them in small clearings to prevent contamination with mortar. The mortar should be less viscous than the pointing mix, as it needs to spread when the stone is placed (not slid or dropped) into position; it will be laid with a rippled surface slightly higher than the required joint so it will not jam up. Pressure on lime mortar rapidly changes it in true Newtonian liquid style from mobile jelly to stiff mastic, as does the suction of porous stone, so don't hang about – get that stone on and positioned!

Now check the stone for true with all means possible, ensuring there is no dogleg when holding a straightedge across the face to the course below.

The next stone should follow on the same bed with a dry perp – some will insist that the mortar should be spread on the perp joint before positioning, but don't be swayed (if they persist let them demonstrate, note the struggle and mess involved, then make your own choice).

Sadly there will be times when all does not go well, with a wall that will not come true due to any number of issues, so be prepared to do some dressing in. Using chisels and light mallet you may need to take down raised edges to pull the surface into *appearing* flat by replicating the tooled finish on the new surface. If you feel tempted to use a spinner, be prepared to do the whole wall to get the same texture everywhere.

BACKFILLING

Except for areas such as parapets, ashlar will be the facing of the wall and will have a backing of rubble walling, block or brickwork that will require stainless steel ties to bind the two walls together, and if no cavity is designed, a rubble infill. The best way to build these is consecutively, filling in the core with clean stone, scalpings or gravel in a mortar matrix up to the levels of the course before laying the next course. Do not over tamp this down as it could dislodge the stone if pushed in too much.

POINTING

It is now the time to fill all low joints up to the level of the face, so prepare a stiff mortar and keep everything clean as before.

What you might notice is that some joints have dry crumbly mortar; this is because the stone was too dry and has sucked the moisture out before the mortar could begin a set. So clean them out to a workable depth and repoint.

Cleaning off fresh mortar with a damp sponge can sometimes result in a toffee coloration. It is best to use clean polished pine to rub off joints as they stiffen up; this compresses as well as finishing nicely.

THE KNACK TO POINTING
1. Load the hawk with very stiff, but malleable mortar, work it well until it is fully mixed and of a stiff putty consistency.
2. Press down with the trowel, and smooth it back to compress at the front nearest the wall to a thickness just less than the joint width.
3. With the back edge of the trowel, cut off enough to fill the joint to the face and push this off the hawk. It should stick out from the trowel in its original shape, if it is the correct consistency – too wet and it will squash up; too dry and it will fall off.
4. Slide the mortar into the joint and push to the back with the edge of the blade, ensuring it does not catch on the edge.
5. Twist the trowel back and slide up to cut the mortar off with the top arris of the joint.

Pointing.

CHAPTER 14

HONEST REPAIRS

The honest way to repair masonry is with good skills, quality craftsmanship and the ability to please the eye.

Many repair projects come under the aegis of learned individuals who tend not to be aware of the skills available, or levels of quality achievable by the artisan, and impose their own character into the work. Probably the worst time for such repairs was during the nineteenth century, when eminent architects such as Gilbert-Scott, whilst being solicitous over the condition of the building, would bring about such profound changes as to destroy the soul of the original – a favourite method was scraping off internal renders and exposing stone never meant to be shown.

It is possible nowadays to gain qualifications in the conservation of these precious items by attending a weekend course or attending university without ever having any appropriate hands-on experience. The result of such impractical education is often a heightened sense of the student's worth without respect for those with less formal education.

RIGHT: Medieval tooled surfaces, probably using a combination of large boasters and axing. The seemingly coarse finish is worked quite level; it may have had a light render or limewash covering, the surface allowing the material to key in.

OPPOSITE: A mediaeval window in Devon, rejuvenated with sympathetic repairs by the author. The ferramenta had rusted and expanded, causing the stones to split; this was removed and the iron ends replaced with stainless steel, while the fragile structure was pinned together discreetly (discernible only to the trained eye), missing areas were rebuilt using lime mortar and Dutchmen where possible. A subtle sheltercoat blended old and new together, allowing the patina of time a good start.

These buildings were built by artisans and should be maintained and repaired in the same manner. Unfortunately all too often the approach is determined by what is in fashion at the moment rather than what the building needs. Some examples, good and bad, are shown here; other approaches are scattered through the book.

DIFFERENT APPROACHES TO REPAIR
Alternatives to New Stone in an Old Building

The following pages illustrate how new stone can be used in old buildings with varying degrees of success.

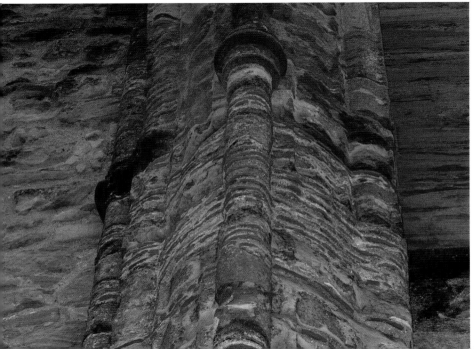

TOP LEFT: Weathered copings repointed after rebedding with added stainless cramps as a practical repair to historic fabric.

TOP RIGHT: A restraint in Ghent that could win an award for 'honesty'! The main benefit is that this is reversible and allows for future work when progress produces better intervention techniques.

LEFT: Contrived: This is the famous tile repair, where the material is termed an honest repair. In some ways it fits with the crude moulding of the weathered original (what is left of it) but, when overused, it reduces the overall effect to one of contrived antiquity.

Quality tile repair: limewashed to harmoniously blend in. The repair is visible but does not offend.

Tidy and efficient repair to missing masonry: The new material is well applied and supports the original visually.

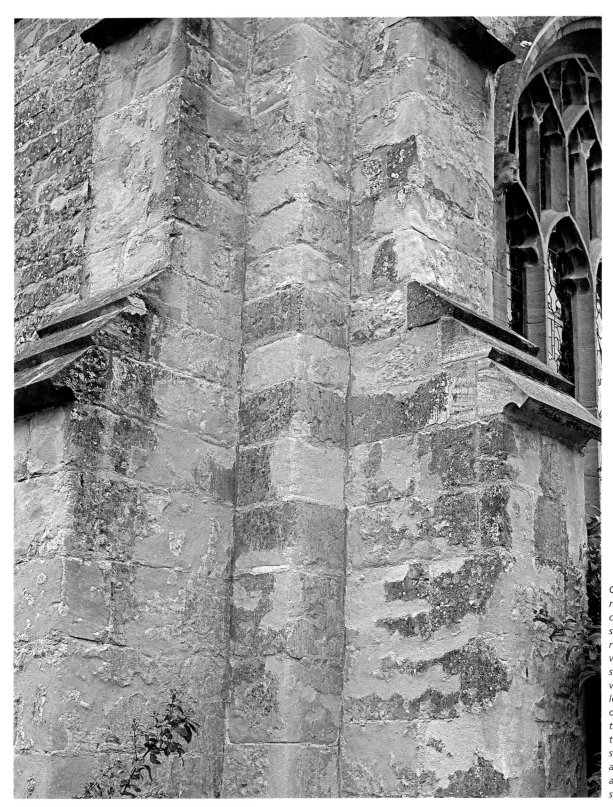

Crude mortar repairs, contoured to supposedly replicate weathered stone: This is wrong on many levels, from the odious texture to the denial of the quality of stonemasonry – an amateurish attempt to be safe.

Considered approach: the surface of the stone on this building, as well as many stones, had deteriorated to the condition where a total rebuild was considered. Instead the whole surface was worked back by hand and machine, and new stone was pieced in with extremely fine joints matching the original with only minimal areas of essential mortar repair – a testament to the intelligence and stonemasonry skills of those carrying out the work.

LEFT: The ashlar of this building was treated with a 'waterproofing' coat. This has caused terrible problems with trapped moisture, as well as giving an unnatural sheen to the wall.

BELOW: An interesting alternative to proper repair is just to bodge! A useful porch is an unusual addition to Sherborne House.

Ill-considered approach: various types of repair using hard cementitious mortars, none of which match the original material and will continue to stand out and wear at a different rate for decades to come.

Coarsely finished surfaces of wall blocks in Wakefield Cathedral: These would originally have been rendered over. The hideous cement pointing and the filthy condition of the stone speak volumes of the care with which these buildings are maintained.

RESIN ADVICE

To glue bits of stone together traditionally stonemasons use polyester resin with catalyst. This can be purchased as a liquid or paste, clear or coloured, and it is always useful to have a tin in the workshop.

It will not work on damp surfaces, and if there is a lot of dust it will stick to that and not the stone.

It is always the case that a surplus of resin is used, so do not wipe the resin off as it squeezes out as this will soak into the stone around, leaving an indelible stain; let it cure and it will break off easily.

Likewise always let resin cure before cleaning it off tools and porous surfaces; this includes any you get on your skin – do not smudge it around.

There are other resins available, such as epoxides and acrylics. They all have their advantages, and it is up to the requirements of the application and what the performance should be.

Super-glue is a cyano-acrylate that is useful for quickly attaching bits into place for more permanent attachment to be carried out subsequently.

Be aware of the safety and handling issues when using any organic chemicals. Always have to hand the Material Safety Data Sheet (MSDM) or Care of Substances Hazardous to Health (COSHH) information, available from the suppliers or on the internet.

MAKING REPAIRS

The only real problem that stonemasons have is that the stone breaks – usually after considerable work has been put in; this can also include large spalls from overenthusiastic pitching or chisel work. The following basic repair techniques should suffice at this level of expertise, but do not cover the use of mortars, reversible interventions or piecing in.

Spalls

Re-attaching a small piece of stone into a depression that it came from is relatively easy. Clean all dust off the surfaces. On a clean surface mix a small amount of polyester resin stone adhesive, making sure no dust or fragments are included. Then lightly butter up the two surfaces and squeeze them together. The aim here is to get a very fine joint across the whole surface, so it is completely attached. Once cured, the stone can be worked as normal, and the

repair should have such a fine line that it appears to be an inclusion.

Fixing Broken Stone

Techniques for repairing broken stone will always depend on the size of the piece and the quality of the stone. Dense fine grained freestone will break into pieces that can be married together without loss of material, whilst softer coarser grained stones can crumble around the break making discrete repairs more of a task.

If the stone is only in two pieces, then the first thing to do is get them matched up and held in position while dowel holes are drilled. Brush or blow all loose particles off the surfaces and put the pieces of the stone together, secured by a ratchet strap or rope tightened by a windlass.

Then on a side that will not be seen or worked, drill holes that pass across the break into sound stone and clean them free of dust. Place a collar around the hole made from a piece

DRILLING ACROSS THE BREAK

Drilling through the outer surface across a break is always the preferred method for the following reasons:

The drill holes are accurately placed on both pieces of the stone.

The time limit of setting resin does not affect the positioning of the stone.

As a courtesy to future stonemasons who may work on this stone – this ensures that the dowel's presence can be more easily predicted (cutting through stone and discovering hidden metalwork is not fun).

Unless there is absolute certainty that the stone will not disintegrate through vibration, try not to use percussive drills; drill bits can be sharpened on the workshop grinder to keep them keen against the dulling of the stone.

As you drill through stone the hole will be small (as the bit) but on the other side it will probably spall out as a crater; these bits must be cleaned out as they will jam up the marrying of the stone. The void will be filled with resin giving a dovetail key to the fix.

of gaffer or masking tape to let any spill/overflow sit on the stone without sticking to it. If possible it is best to either pour the resin or inject it into the hole as this will ensure the resin gets right down; trying to get a paste into a deep, narrow hole using a spatula is very difficult due to its stickiness and also the race against (setting) time.

With the resin in the hole, insert the dowel completely (push it in with another piece of dowel) so that it is encased completely in resin, and leave to cure; the resistance of the resin can require some pressure on the dowel, which if the piece was not strapped together would be pushed out of place.

The stone is now ready to be used.

Fixing Small Pieces

If the piece broken off is to be fixed on the face of the stone (a possible scenario is of a nosing coming adrift during fixing) then it may be impractical to fit a dowel right through, or even use one at all. It is not advisable to attach stone to other stone just by a joint of resin, as while the resin is strong, the stone either side of it could come under stress from weather, heat or water issues and possibly drop off.

So this is what you can do. Drill a short hole into the main stone and very carefully and gently make a depression in the small piece that corresponds to the other; use a chisel to grind this in and put some undercut in to help the grip.

Now cut a dowel that just projects past the face of the (large) stone. Place it and offer up the fragment to check it fits flush; make any adjustments to the holes as necessary. Devise a method to hold the stone in place while it cures, that you can fit one hand into while you hold the fragment in place.

Place the resin into the dowel holes, ensuring they are full by poking and squeezing, and put the dowel in. Then squeeze the fragment into place (leaving oozing resin to cure) and lodge it into place using your bespoke method.

THE BEST FINISH

But what if the original is eroded to the point that it has no tooling visible or it has all disappeared completely?

Unless it is specified as a rubbed or polished finish the stone should be left with a fine tooled surface of neat and regular chisel cuts all at an appropriate angle (this may depend on the texture of the stone). This is how it should be done. People with only a theoretical interest in stonemasonry will often call this an 'honest repair', which whilst sounding noble indicates that it is only because of this new piece of stone that they can recognize the repair – and often they will (erroneously) consider it poor workmanship.

This way of working is necessary to prolong the life of a building and stone, and has licence in the structure's history. Sadly it is often grouped incorrectly with one of many 'artistic' or 'museum' styles of repairs carried out in the name of restoration and conservation by well-meaning but ineffectual (in terms of practical knowledge) amateurs, who consider that the necessary skills to maintain buildings do not exist anymore and therefore the correct method should not be attempted.

A traditional building needing replacement stones should be worked on by trained artisans competent in the proce-

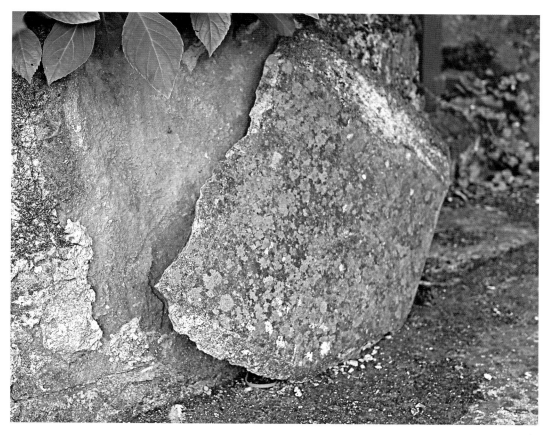

Faults or vents in stone can show up years after use.

dures of traditional stonemasonry, just as conservators undertaking this type of work should have skill and training equivalence. If you have worked through this book, well on your way to acquiring these ancient – and ostensibly successful – ways of working, let me welcome you to one of the most satisfying professions in the world.

ABOVE: This Roman wall, with its various historic repairs, is crude but strong, and illustrates the timelessness of good masonry.

LEFT: The endurance of good masonry. Even after explosives were used here during the Civil War, large chunks of wall still held together. Note the massiveness of the core.

GLOSSARY

arris Projecting edge where two worked surfaces meet that is not a mitre or cusp – basically the outer edges of a block (a widely used term in masonry which should be bandied about whenever discussing stone).

ashlar Squared stone laid in regular height courses, with even joints.

banker Stand for working stone on.

bed Horizontal surface of a stone, or the flat horizontal place in a structure where it is to be set.

bedmould A full size drawing that shows all the details of the stone from above (in technical drawing this would be the plan), including hidden detail.

bevel Sloping surface.

boasted Surface finish of fine regularly cut lines.

boaster Wide chisel used for final dressing of stone surface.

bonding Overlapping of stones so that vertical joints are staggered, allowing the wall to lock itself in place (to check the veracity of this, build walls from lego blocks in both styles and see which is the strongest).

bone in To produce levels on a stone prior to working a flat surface.

box trammel Instrument to scribe in a line parallel to an edge.

bulking Sand that has been allowed to get damp to the point where its volume actually increases, due to the water sticking the grains of sand together; to compensate add more water and the grains will slide together reducing the volume.

centring Frame to support arch during construction.

chamfer Surface angled to join two flat surfaces.

check A small level surface cut in during working of the stone.

chevron V shaped mould

clastic Stone that has other stones or materials in its matrix.

corbelling A piece of stone projecting from the face of a wall, generally used to support a cornice or arch.

coxcomb Curved stone drag, used for concave mouldings.

cramp Method of tying two stones together across the joint.

cusp Decorative projecting point of an inner curve in tracery.

deadeye Triangular depression cut in the flat areas of tracery, which does not go right through the stone.

details Stone components of a building which are not walling – generally masonry that has been worked to a templet or mould.

diagenetic Change that occurs after stone has been formed.

differential weathering The erosion of materials in the same place, where the rate of loss varies according to the durability of the material.

dogleg A continuous line that has one deep bend.

draft A narrow worked area to set a height.

dressed Stone that has been dressed has been worked.

droved work Work made with a wider chisel (2.5in 65mm) without attempting to keep the (cut) lines continuous.

DPC (damp-proof course) A barrier in a wall, which is level in structure and usually just above ground level, using a material that prevents the travel of moisture.

entasis Concave curve added to a shaft to prevent it looking too waisted

extrados Outside line of an arch.

face-mould A templet that shows the shape of the stone from the front.

fettling Finishing off stones after they have been fixed, to blend them in with surrounding stones or to bring moulding to a level.

fillet Small mould of square section.

finial Ornament at the top of a detail, finishing off an upright.

fixing Building stones into a structure.

foil Opening between cusps.

free-stone Stone that is regular throughout and can be worked in any direction.

grout Liquid mortar for filling joints and voids in masonry.

guesstimate A figure arrived at without calculation.

hawk Board held in the hand on which mortar is kept whilst pointing.

HSS (high speed steel) Describes drill bits that do not have TCT as a cutting edge.

intrados Inside line of an arch; often termed the soffit, which correctly applies to any surface overhead in openings or similar.

jamb Vertical side to an opening.

joggle Indent or channel cut in joint section of stone.

joint The join between two stones that are put together.

keying A roughened or scored surface to provide grip or purchase.

keystone Top stone of an arch, locking into place the structure.

lancet arch Commonly known as gothic, where the two centres for the arcs start on the springing line.

lift The working level, or platform, on a scaffold.

lintel Stone spanning an opening, supporting the masonry above.

lozenge Diamond shape in masonry and heraldry.

mitre The line where two similar mouldings meet, either internally (going into a corner) or externally (forming a projecting corner).

monial Where a fillet follows a curve in tracery to form the surface at the face of the foil at the same height as the fillet to the frame.

monolithic Made of one stone, used to indicate something massive.

mould Alternative term for templet; some stonemasons use one term only or may insist on specific definitions, but everyone understands what is meant.

moulding The external shape of a worked piece of stone.

mullion Upright divider of a window.

ogee Moulding of two (reversed) curves.

ovolo Moulding of a quadrant (quarter circle).

perp Upright joint in masonry.

pitch A strike in stone to cause a directed fracture.

pitcher Heavy chisel without a cutting edge for causing fractures off a line (US: hand set).

plaster Wall covering of mortar.

plinth Base course or block of wall or column.

pyroclastic Caused by superheated materials under pressure; a geology term.

radiating joints Joints that go towards a common centre.

relieved Describes stone that has been cut away, before a more precise cut is made, to prevent spalling or to allow the subsequent cut to be more efficient.

relieving cut Furrow to stop another cut in stone travelling too far.

return The point where a moulding changes direction.

roll A curved moulding, usually circular in section.

rusticated Describes a joint which is recessed back from the face; it may be worked in any style.

saw-lash Marks left on a stone from being cut by saw.

sawyer Producer of stone cut on a saw.

scribe To score a line in stone.

scriber Steel or TCT stylus for marking stone.

section The shape formed when cutting through a stone.

serif Line at the end of a letter, used for embellishment in certain fonts.

setting out Drawing the designs for the stone to be worked to, either at scale or full size.

socle A base for column, finial or urn.

soffit Underside of an arch, lintel, ceiling or overhang.

softening Packing material that is softer than stone, used to prevent damage when storing or transporting.

spall A small piece of stone that has been knocked out, or off, a worked piece.

span The distance of an opening from side to side.

springing line Horizontal level at which an arch starts

SSS (six sides sawn)   Description of stone when ordering from the sawyer/supplier; it will come as a block with all sides sawn, to the dimensions requested.

stop   The end of a run of moulding, which can be into a flat surface or as a termination of a hood (or label) mould around an opening, occasionally having a carved piece on the end (label-stop).

string course   Moulded course of stone that runs horizontally.

stucco   Another term for render.

tallion   French stone axe.

tangent   Straight line that touches a curve without cutting it.

templet   A sheet cut to the shape of a stone on a particular plane, generally though the joint or side of the stone to show the mouldings and details not shown on the bed-mould, but interchangeable with mould as a term.

tooled work   Work finished off by using a chisel in a regular style and running the cuts in continuous lines across the stone.

tracery   Geometric shapes applied to the stone in the area inside the frame of windows and panels.

trammel   Instrument to draw curves.

transom   Horizontal divider of a window.

vault   Arched roof or ceiling.

vernacular   Indigenous building style that uses local materials and traditional methods of construction and ornament, as distinct from historical architectural styles.

voussoir   One stone of an arch.

IN CONCLUSION

Well here we are at the ultimate section of the book, but this shouldn't be end; rather I hope that this volume has whetted your appetite, and given you a taste of the satisfaction and joy of working with stone.

The whole world of masonry beckons and it is now your choice on which road to take, for there are myriad aspects of working stone and, just as a building is made of many components, all contribute to the built heritage.

EDUCATION AND TRAINING

It is never a waste of time to undertake a course of learning, for every tutor will impart their own knowledge in a unique manner and the fellow students met will become friends and colleagues, in all probability for the rest of your life. So do some research and decide whether this is the way to advance and most specifically where are you going to study.

There are colleges that deal with proper stonemasonry throughout the world and these can be found by contacting the regional stone organization of a country, searching the Internet or by word of mouth from stonemasons met. It would be impossible to list them all here, as well as all the vocational and specialized courses available that are run on a smaller scale, so get out there and find out what's available.

WORK-BASED EXPERIENCE

The construction industry is by nature a consistent hirer of temporary workforces as contracts come in that need larger numbers for short periods of time, or small companies need extra hands to fulfil a full order book. So the opportunities to get on site and work with experienced stonemasons will

always present themselves to motivated individuals; this is time well spent and will build confidence in your own skills, as well as being an eye-opener into the ways of the builders' world.

Small workshops can be approached and, because stonemasons are usually the nicest of people, an enquiry can result in an ad hoc internship as a helper or student. Don't be shy in offering your services, but do be humble in your approach – hubris is not a tool to be left holding in an extreme situation.

STONE CARVING AND LETTERCUTTING

The wish to specialize in one of the above is what drives many people to take up stonemasonry; and it is a wise step, for with the skills of a stonemason, the debutant in carving is on a good foundation. A feel for design and the ability to translate this into drawing and three dimensions is obviously an essential part of the process. Some equate the freehand carving of stone as sculpture; if the creation of unique art, as distinct from architectural and decorative carving, is what you aim for then at least proficient skills in the working of the material is going to be very useful.

RESOURCES

The availability of information in this day and age is, possibly, too easy and there is often the chance that when a search for a technique on the Internet results in millions of options, it may be difficult to decide on the right choice; so be cautious and don't be afraid to ask questions of other stonemasons (possibly on one of the many crafts forums that now exist).

Books are tangible, usually written with some authority and, in this traditional craft, don't really go out of fashion, so a small library is always useful and wonderful to own – an advantage with books is they can be taken into a workshop and won't break down if they get dusty!

There are publications covering stone in all its uses and forms, these will usually have an online presence, and can be a good source of information and contacts.

Books You Should Own

Obviously this book is essential, so if you're reading it in a bookshop get to the counter and put the money down straight away!

The following are still available and essential to have:

Fletcher, Banister, *A History of Architecture* (Architectural Press, 1996)
Fleming, John, *et al.*, *The Penguin Dictionary of Architecture* (Penguin, 1991)
Grasby, Richard, *Letter Cutting in Stone – a workbook* (University of Pennsylvania Press, 1990)

Perkins, Tom, *The Art of Letter Carving in Stone* (Crowood, 2007)
Warland, Edmund George, *Modern Practical Masonry* (Donhead Publishing, 2006)

Additionally it is well worth acquiring any good textbook on geometry as taught at school. Many books on construction methods have been published and can be handy for how to repair buildings in original manner. Also look out for local architectural history books about your area, which will show the development of buildings. Similarly local geological surveys will help in identifying stone.

AND FINALLY...

There is also the dedicated world of conservation and restoration, which takes involvement to a different level and can lead on to all manner of interesting projects; this is a subject dear to my heart as I firmly believe that those involved in the preservation of the built heritage should be accomplished in the tools skills and techniques with which it was built.

INDEX

OTHER CRAFT TECHNIQUE BOOKS FROM CROWOOD

Abbott, Kathy *Bookbinding*
Adcock, Sandra *Commercial Floristry – designs and techniques*
Blishen, Nick *Acoustic Guitar Making*
Brooks, Nick *Mouldmaking and Casting*
Brooks, Nick *Advanced Mouldmaking and Casting*
Burke, Ed *Glass Blowing*
Ellen, Alison *Hand Knitting*
Ellen, Alison *Knitting – colour, structure and design*
Fish, June *Designing and Printing Textiles*
Goodwin, Elaine M. *The Human Form in Mosaic*
Grenier, Fleur *Pewter – designs and techniques*
Hunter, Andrea *Creating Felt Pictures*
Parkinson, Peter *The Artist Blacksmith*
Perkins, Tom *The Art of Letter Carving in Stone*
Richards, Ann *Weaving Textiles That Shape Themselves*
Seymour, Martin *Clinker Boat Building*
Smith, Alan *Etching*
Taylor, Chris *Leatherwork – a practical guide*
Thaddeus, Martin, and Thaddeus, Ed *Welding*
Tregidgo, Jan *Torchon Lacemaking – a step-by-step guide*
Waller, Jane *Knitting Fashions of the 1940s*
Watkins-Baker Helga *Kiln Forming Glass*
Werge-Hartley, Jeanne *Enamelling on Precious Metals*